Proceedings (Trudy) of the P. N. Lebedev Physics Institute

Volume 99

Stimulated
Raman Scattering

Edited by
N. G. Basov

P. N. Lebedev Physics Institute
Academy of Sciences of the USSR
Moscow, USSR

Translated from Russian by
J. George Adashko

SPRINGER SCIENCE+BUSINESS MEDIA, LLC

Library of Congress Cataloging in Publication Data

Vynuzhdennoe kombinafsionnoe rasseîanie sveta. English.
   Stimulated Raman scattering.

   (Proceedings (Trudy) of the P. N. Lebedev Physics Institute; v. 99)
   Translation of: Vynuzhdennoe kombinafsionnoe rasseîanie sveta.
   Includes bibliographical references.
   1. Raman effect—Addresses, essays, lectures. I. Basov, N. G. (Nikolaĭ
Gennadievich), 1922–   . II. Adashko, J. George. III. Title. IV. Series.
QC1.A4114 vol. 99 [QC454.R36] 530s   [535.8′46]   82-18238
ISBN 978-1-4757-1407-4   ISBN 978-1-4757-1405-0 (eBook)
DOI 10.1007/978-1-4757-1405-0

The original Russian text was published by Nauka Press in Moscow
in 1977 for the Academy of Sciences of the USSR as Volume 99 of the
Proceedings of the P. N. Lebedev Physics Institute. This Translation
is published under an agreement with VAAP, the Copyright Agency of the USSR.

STIMULATED RAMAN SCATTERING

VYNUZHDENNOE KOMBINATSIONNOE RASSEYANIE SVETA

ВЫНУЖДЕННОЕ КОМБИНАЦИОННОЕ РАССЕЯНИЕ СВЕТА

The Lebedev Physics Institute Series

Editors: Academicians D. V. Skobel'tsyn and N. G. Basov

P. N. Lebedev Physics Institute, Academy of Sciences of the USSR

Recent Volumes in this Series

PREFACE

Results are presented of an experimental investigation of the angular distribution of the near-axis emission of various components of stimulated Raman scattering which are characterized by a strong forward directivity (repetition effect). "Rings of second class" in off-axis stimulated Raman scattering emission are investigated.

Experiments that have revealed the existence of a new nonlinear-optics phenomenon, namely self-focusing of stimulated Raman scattering, are described.

Results of an investigation of the spectral distribution of the stimulated Raman emission components are reported.

One of the articles is devoted to a theoretical treatment of the angular distribution of stimulated Raman scattering.

All the articles are trail-blazing in their fields. The anthology systematically explains many aspects of the present state of the study of stimulated Raman scattering.

CONTENTS

INVESTIGATION OF THE ANGULAR DISTRIBUTION OF
STIMULATED RAMAN SCATTERING OF LIGHT

A. N. Arbatskaya

The angular distribution of the paraxial emission of the first, second, third, and fourth Stokes components of stimulated Raman scattering of light (SRS) in organic liquids (benzene, carbon disulfide, nitrobenzene) is investigated. A "duplication effect," consisting in the fact that the angular distribution of the paraxial radiation of all the Stokes components duplicates the angular distribution of the exciting radiation emerging from the scattering medium, is observed. Off-axis class-II SRS emission in the same liquids is investigated. A new interpretation is proposed for the off-axis class-II emission on the basis of the Stokes − anti-Stokes four-photon processes. An estimate is presented of the relative intensity of the off-axis SRS radiation and of the total intensity of the second and higher Stokes components.

INTRODUCTION

Little more than a decade has elapsed since the discovery of stimulated Raman scattering (SRS) of light, but hundreds of papers have already been devoted to this phenomenon. The interest in SRS is fully justified by the variety of the laws that govern it. It should be noted although SRS has been intensively studied, there are still many unclear aspects, both as a result of the complexity of the phenomenon itself and because of its connection with other nonlinear processes that occur when intense optical radiation interacts with matter.

The investigations of the angular distribution of SRS were initiated shortly after its discovery, especially after experiment has revealed radiation at anti-Stokes and higher Stokes frequencies, propagating along the generators of cones with axes along the direction of the exciting radiation. The apex angles of these cones were measured for a number of substances under various SRS excitation conditions. As to the other characteristics of off-axis SRS radiation, including the intensity, they remained practically uninvestigated to this day.

Photography of SRS radiation at anti-Stokes and higher Stokes frequencies reveals usually several rings and a central spot corresponding to the SRS in the direction of the axis of the exciting light. At the first Stokes frequency there is only a central spot (and an absorption ring). Since principal attention was paid to the study of the rings, the angular distribution of the SRS radiation in the central spot remained uninvestigated. Yet for a number of reasons one should expect this distribution to be unusually narrow, only slightly exceeding in width the angular distribution of the exciting radiation. Interest in this question has increased further because the angular distribution at the first Stokes frequency can be used to assess the role of coherent processes and of SRS phenomenon. Therefore the first task of the present investigation was a detailed study of the angular distribution of the first Stokes SRS component in a number of objects under various excitation conditions.

The second task was the study of the angular distribution of the off-axis radiation and a comparison of the obtained data with the available theoretical concepts concerning the processes that occur in the SRS phenomenon. The appearance of the off-axis radiation cones was explained soon after Townes's discovery of this radiation. It became clear subsequently that this explanation is suitable only for part of the observed cones, which were called cones of class I. It was also made clear that SRS is subject to definite perturbing action by other nonlinear optical phenomena, such as self-focusing of the light in the medium. The appearance of off-axis radiation cones, which do not agree with the theory mentioned above (they were called cones of class II), was attributed to self-focusing of the light. It was assumed that thin self-focusing filaments are produced in the medium, and it is in these filaments that the SRS radiation propagating along the class-II cones is excited.

The foregoing treatment of class-II radiation cannot be regarded at present as incontrovertible. According to present-day concepts, the onset of thin self-focusing filaments in a medium in which high-intensity laser radiation propagates does not correspond to the real processes that occur in the medium. We note also that off-axis class-II radiation was observed in some experiments in the absence of self-focusing. It is therefore of great interest to ascertain the causes of the class-II radiation.

Another independent problem is the study of the relative intensity of the radiation propagating along the axis and of the off-axis radiation for different SRS components, as functions of the character of the scattering objects and of the excitation conditions. We note that no research was carried out in this direction.

The study of the relative intensities of the rings and of the central spot is important from the point of view of explaining the mechanism of formation of various SRS components. Up to now, the measured distributions of the intensities among the various SRS components were the summary intensities of the radiation contained in one component or another. Yet the mechanisms that lead to the appearance of cones of different type and of the central spot can in principle be entirely different. Therefore separate measurements of the intensity of each of these formations can serve as a basis for the understanding of the most important SRS mechanisms.

CHAPTER I

SURVEY OF THE LITERATURE

1. Experimental Investigations of the Angular Distribution of SRS

Much research was done by now on SRS. Soviet and foreign physicists developed the classical and quantum theory of the phenomenon. Numerous experimental studies were made of the SRS spectra of liquids, solids, and gases excited by various methods, to the distribution of the intensities in the SRS spectra, to the energy characteristics and thresholds of the SRS, the waveforms and durations of the SRS component pulses, the coefficients of conversion of the laser energy into SRS-component energy, and to the angular characteristics of the SRS components. The results of these investigations are reflected in a large number of original papers, as well as in reviews and monographs [1-9].

SRS finds practical applications in Raman lasers and amplifiers. The first liquid-nitrogen Raman laser, in which it became possible to exceed the brightness of the pump source, was developed by Grasyuk and co-workers [10].

Principal attention is paid in the present review to the experimental and theoretical investigations of the angular characteristics of SRS.

The SRS phenomenon was first observed in 1962 by Woodbury and Ng [11] in an investigation of the emission of a ruby laser Q-switched by a Kerr cell with nitrobenzene. It was observed that the emission of this laser contained, besides the ruby emission at the fundamental frequency, intense radiation in the infrared region. Hellwarth [12] correctly interpreted this effect as SRS. The communication [11] was soon followed by investigations of the SRS of various liquids placed in the optical cavity of a Q-switched ruby laser, where a "giant pulse" was produced. Eckhardt et al. [13], followed by Geller et al. [14], investigated the SRS of several liquids (benzene, toluené, pyridine, cyclohexane, nitrobenzene, and others). One or two Stokes frequencies were observed in the spectrum of each of the substances.

It was subsequently observed that SRS is produced when the radiation of a giant laser pulse is focused into a medium placed outside the laser cavity. Using an installation of this type, Terhune [15] registered not only Stokes but also anti-Stokes Raman frequencies. It turned out that the anti-Stokes radiation leaves the cell not in the direction of the excited radiation (as does the Stokes component), but at a certain angle — on the surface of the cone whose axis coincides with the direction of the excited radiation. Each harmonic has has its own angle of inclination of the cone generator, so that concentric rings of different colors are produced on a color film placed behind the cell with the investigated liquid and perpendicular to its axis. Intense anti-Stokes scattering of liquefied hydrogen, oxygen, and nitrogen was observed by Stoicheff [16]. Later on, SRS spectra were obtained for gases [17, 18] and solids, diamond single crystals, calcite, and sulfur [19].

The unusual character of the angular distribution of the SRS, observed in [15, 16], has attracted the attention of many workers. Following the first experimental studies, a theoretical investigation was made of the questions connected with the appearance and spatial distribution of the SRS components. Townes and co-workers [20, 21] have developed the relations that must be satisfied by the wave vectors of the exciting radiation and of the various components of the scattered radiation (phase-synchronism relations). In the general case these relations take the form

$$k_0 + k_{n-1} = k_{-1} + k_n, \qquad (1.1)$$
$$k_0 + k_{-1} = k_{n-1} + k_{-n}. \qquad (1.2)$$

Here k_n are the wave vectors of the light waves with frequencies

$$\omega_n = \omega_0 \stackrel{+}{-} n\omega_v, \qquad (1.3)$$

where $n = 1, 2, 3, \ldots$; the minus sign corresponds to the Stokes frequencies, and the plus sign to the anti-Stokes frequencies; ω_v is the vibrational frequency of the molecule.

In the simplest case at $n = 1$ we have

$$2k_0 = k_{-1} + k_1. \qquad (1.4)$$

It follows from (1.4) that anti-Stokes radiation is emitted only in a direction that forms a cone with an angle θ_1 to the direction of the exciting radiation. For small angles, calculations yield

$$\theta_1^2 = \frac{1}{n} \frac{\omega_0 - \omega_2}{\omega_0 + \omega_2} \left[\Delta n_1 - \Delta n_{-1} + \frac{\omega_2}{\omega_0} (\Delta n_1 + \Delta n_{-1}) \right]. \qquad (1.5)$$

Here n is the refractive index of the medium for the frequency ω_0; Δn_1 and Δn_{-1} are the differences of the refractive indices for the frequencies $\omega_0 \pm \omega_2$ and ω_0, respectively. For normal dispersion of the medium, all these quantities are positive. Since $\omega_2 / \omega_0 \ll 1$, we get

$$\theta_1^2 \cong \frac{1}{n} \frac{\omega_0 - \omega_2}{\omega_0 + \omega_2} (\Delta n_1 - \Delta n_{-1}). \qquad (1.6)$$

It follows therefore that the angle θ_1 is determined by the slope of the dispersion curve of the medium.

Relations (1.1) and (1.2) for the wave vectors were experimentally confirmed for the SRS components in calcite crystals. Chiao and Stoicheff [22] were the first to observe four anti-Stokes and the second Stokes components in the SRS spectrum of calcite, and measured the apex angles of the emission cones. An absorption ring at the first Stokes frequency was also observed.

In the cited reference, the scattered radiation was photographed directly on a photographic plate placed in the path of the beam emerging from the crystal. If the beam of the exciting radiation is perpendicular to the entrance and exit faces of the crystal, the anti-Stokes components produce on the photographic plate dark concentric rings. By measuring the distance from the photographic plate to the crystal and the diameters of the rings produced on the plate, the angles at which the radiation diverges and the positions of the vertices of the cones were calculated. It turned out that the position of the vertex of the emission cone depends on the position of the focus of the focusing lens relative to the crystal, and a strong dependence of the angles on the focal length of this lens was observed. The authors have therefore extrapolated the data for angles obtained by focusing the exciting radiation with lenses of different focal lengths (8, 20, 30, 50, 127 cm) to the ideal case ($f = \infty$). In this case the pump wave can be regarded as plane.

The SRS-component emission angles calculated from the phase-synchronism relations (1.1) and (1.2) are in much worse agreement with experiment when the angular distribution of SRS in liquids is observed. Thus, for example, Hellwarth and co-workers [23] obtained, in an investigation of the SRS spectra in nitrobenzene, a considerable discrepancy between the experimental results and the calculations.

The calculation of the angles yielded the values $\theta_1 = 2.5 \pm 0.2°$; $\theta_{-2} = 12.4 \pm 1°$; $\theta_{-3} = 16$ or $12 \pm 1°$. Experiment yielded substantially different results: $\theta_1 = 3.1 \pm 0.1°$; $\theta_{-2} = 3.9 \pm 0.1°$; $\theta_{-3} = 3.8 \pm 0.1°$.

Further investigations of the angular distribution of SRS in liquids have shown this phenomenon to be more complicated than it appeared to be at first. Garmire [24] has established that two classes of SRS of higher order exist in liquids. Class-I radiation produces an angular distribution that satisfies the synchronism conditions (1.1) and (1.2). Class-II radiation is emitted along generators of cones whose angles differ from those calculated from the synchronism conditions. In the case of class-II radiation the rings can be diffuse and can sometimes split into several components.

The onset of class-II radiation is attributed in [24] to the breakup of the laser beam in the liquid into very thin filaments (see Section 2).

A prominent place is occupied in Garmire's research [24, 25] by the question of the role of the first Stokes component upon excitation of anti-Stokes components. According to Garmire, anti-Stokes components of class I can be emitted in the presence of a sufficiently intense first Stokes component at the synchronism angle. If there is no emission in this direction, then only class-II radiation can be excited.

In liquids there is usually not enough intense radiation at the Stokes frequency in the direction of the angles corresponding to the phase-synchronism conditions, owing to the strong forward directivity of the radiation at the first Stokes frequency. According to estimates [25], the intensity of the first Stokes component in benzene, in a cell 10 cm long and at the angle of phase synchronism, is 10^{-6} of the intensity of this radiation in the direction of the exciting beam. Therefore the radiation of the rings of class I will be very weak and the radiation of class II will mask that of class I. Accordingly, the SRS radiation of higher order in liquids at angles that do not coincide with those obtained from phase synchronism, which was observed in the cited paper [23] and in a number of others [26-29], is in fact class-II radiation.

To observe class-I radiation in liquids, Garmire [25] performed a number of experiments. In these investigations, the SRS was excited by a parallel light beam from a Q-switched ruby laser of ~10 MW power. To increase the intensity of the Stokes radiation near the direction corresponding to the synchronism conditions the axis of the cell with the investigated liquid was mounted at an angle $\simeq \theta$ (the synchronism angle) relative to the direction of the laser beam. The cell was 10 cm long and its windows were strictly parallel. Part of the Stokes radiation at the angle θ was reflected back from the glass—air interface, so that the intensity of the radiation in this direction was increased by 10^4 times. This increase of the intensity of the Stokes radiation was sufficient to excite class-I anti-Stokes components. The first and second class-I anti-Stokes components were observed in the form of intense short arcs. These arcs and the first Stokes component were located on two opposite sides of the laser beam, as expected from the phase-synchronism relations. The radiation of the class-II anti-Stokes components of the first and second orders was observed in these experiments in the form of solid rings. The diameters of the rings and their intensities were independent of the amplification of the first Stokes component.

Anti-Stokes radiation of class I was sometimes observed in organic liquids (benzene, nitrobenzene) in cells of great length (25-50 cm) without special measures to amplify the first Stokes component [25] in multimode excitation. The intensity of this radiation was much less than the intensity of the corresponding class-II rings.

The anti-Stokes components of class I were observed sufficiently easily in acetone and in cyclohexane [25, 30]. It was suggested in [25] that in acetone and cyclohexane the class-II rings are weaker than in benzene and nitrobenzene, and therefore they do not mask the class-I rings.

The class-I radiation observed in calcite is explained in [25] in the following manner: scattering of the light by the inhomogeneities in the calcite is so strong that the intensity of the first Stokes component is appreciable in a large angle interval (including at the phase-synchronism angle), so that intense class-I radiation at anti-Stokes frequencies is produced and masks the weak class-II radiation. Thus, according to Garmire [25], the ratio of the intensities of the class-I and class-II radiation depends on the properties of the scattering medium and on the experimental conditions. In principle, these two classes of radiation are produced independently of each other.

The angular distribution of the SRS in gases was investigated in [3, 17]. In hydrogen gas, the observed angle of the anti-Stokes radiation agrees well with the value calculated from the phase-synchronism conditions. Since the dispersion in hydrogen is small and is proportional to the density, the anti-Stokes radiation of first order should make an angle proportional to $\mathscr{P}^{1/2}$ with the direction of the axis ($\mathscr{P}$ is the pressure of the gas). This conclusion is confirmed by experiment. At low pressures the apex angles of the cones of the anti-Stokes and higher Stokes components are small. Therefore the emission of these SRS components in gases takes place in practice within the limits of the aperture of the exciting-light beam.

Among the liquids in which the angular distribution of the SRS satisfies the phase-synchronism conditions (1.1) and (1.2), notice should be taken of liquid nitrogen. As shown in [31], in liquid nitrogen the calculated angles of the cones of the anti-Stokes and second Stokes components of the SRS are in satisfactory agreement with the values of the angles measured in that study. It should be noted that in [32] it was observed that the apex angles of these SRS-component cones depend not only on the focal length of the lens that focuses the exciting radiation beam in the cell, but also on the length of the cell with the liquid nitrogen.

Mention was already made of the good agreement between the theoretical and experimental data for the apex angles of the cones of the SRS components in calcites. It was shown in a group of studies [33-38] that the angular distribution of SRS in crystals is much more complicated than in isotropic media. In addition to the previously observed cones of SRS emission in crystals, identical with the SRS cones in isotropic media (principal radiation), an additional radiation was observed. The existence of this supplementary radiation was theoretically predicted by Lugovoi [35].

In a uniaxial calcite crystal, in accordance with the theory, supplementary-radiation cones were observed in the form of surfaces of revolution [33, 34, 38]. In these experiments, a linearly polarized ruby-laser pump beam passed through the calcite crystal as the ordinary wave. The rings of the principal and supplementary radiation were observed for a number of anti-Stokes and second Stokes components of the SRS. The principal radiation rings do not depend on the orientation of the crystal relative to the laser beam, and the centers of these rings always remain on the laser-beam axis. The supplementary-radiation rings depend on the angle ν between the direction of the optical axis of the crystal and the wave vector of the pump beam. The centers of the supplementary rings are shifted relative to the center of the principal rings in the direction of the rotation of the optical axis of the crystal.

The dependence of the supplementary-radiation rings on the angle is the following. At small ν, the principal and supplementary rings of a given SRS component are first split apart. On further increase of ν, the shift of the center of the supplementary rings increases, as does the ring diameter. The supplementary radiation is observed only in a limited interval of angles ν.

In [37, 38] they investigated the angular distribution of SRS in the biaxial crystal aragonite. In contrast to the uniaxial crystals, the principal and supplementary radiation of the SRS components in biaxial crystals propagates along cones of revolution only at a definite orientation of the crystal and only at small values of the angle ψ between the direction of one of the optical axis of the crystal and the wave vector of the pump beam. When these conditions are not satisfied, the cones of the SRS are generally speaking not surfaces of revolution. Therefore the photographed radiation rings are not circles.

In [39, 40] they investigated the angular distribution of SRS in diamond (frequency 1332 cm^{-1}). The diamond crystal was a plate 2.2 mm thick. Radiation cones of the first, second, and third anti-Stokes components and of the second Stokes components of the SRS were observed, as well as minima of the intensity in the diffuse scattering of the first Stokes component.

The observed cone angles are in good agreement with the theory based on the phase-synchronism condition obtained by Townes [20, 21]. It turned out that the angles of the cones do not depend on the power of the exciting radiation, but do depend on the convergence angle of the exciting radiation. Rivoire [41, 42] investigated the spatial distribution of SRS radiation in benzene, nitrobenzene, and carbon disulfide. In these investigations, besides the photographic method, a photoelectric method of recording the SRS components was used. The radiation emerging from the cell with the investigated liquid was directed with a total-internal-reflection prism onto the slit of a spectrograph. The total-internal-reflection prism was mounted on the stage of a goniometer. By rotating the stage it was possible to project on the slit of the spectrograph various sections of the scattered-beam cross sections and to study its structure by a photoelectric method. The use of this method, naturally, calls for high stability of the SRS excitation, otherwise a large scatter is obtained in the intensities of the spatial distribution of the SRS. Rivoire obtained a complicated structure of the spatial distribution of the radiation at the first Stokes component and the other SRS component, exceeding the limits of the measurement errors.

2. Influence of Self-Focusing of the Light in the Scattering
Medium on the Angular Distribution of the SRS Components

When intense light waves propagate through a material medium, an important role is played by effects of self-action of the waves, effects that frequently determine the course of the nonlinear processes in the medium. Within the framework of nonlinear optics, the effects of self-action of intense light waves is described by the real part of the nonlinear susceptibility χ_{ijk}, χ_{ijkl}, etc.

In a strong optical field, the refractive index, just as the susceptibility, depends on the field intensity E. For transparent media this dependence is quadratic,

$$n = n_0 + n_2 \, |E|^2. \tag{2.1}$$

Because of the nonlinear dependence of the refractive index on the field intensity, a number of phenomena occur, and the most interesting from our point of view is the self-focusing of the light. The causes of the nonlinear part of the refractive index can vary. The most important of them are the change of the polarizability and the change of the density of the medium in the strong optical field. The change of the polarizability is due to the high-frequency Kerr effect [43-45], to deformation of the electron shells [45, 46], or to the increase of the polarizability of the excited molecules [47, 48]. The change of the density of the medium may be due to electrostrictions [49]. Generally speaking, the nonlinear increment of the refractive index can be due to all the foregoing causes. Usually, however, it is possible to separate the most significant mechanism that produces this increment.

In the high-frequency Kerr effect and electrostriction, the regions of maximum light intensity are usually simultaneously also the ones with the highest optical density. Therefore the nonlinearity of the refractive index

5

leads to a concentration of the energy, i.e., the beams contract from the periphery into a region where the field is maximal. This effect was named self-focusing of the light beam.

An effect counteracting the concentration of the optical energy is diffraction. There should exist consequently a certain critical density, starting with which the focusing effect can exceed the diffraction divergence of the beam. The critical power is equal to [43, 50-53]

$$P_{cr} = (1.22 \lambda)^2 C/256\, n_2, \tag{2.2}$$

where λ is the laser emission wavelengths; n_2 is the change of the refractive index in a constant electric field, and is connected with the Kerr constant B by the relation

$$n_2 = 2\lambda B/3. \tag{2.3}$$

When the power P exceeds the critical value, self-focusing sets in. It was believed for a long time that self-focusing causes the laser beam to split into thin filaments with high light intensity in each, and these filaments form unique optical waveguides.

Lugovoi, Prokhorov, and others have developed a different model of self-focusing of light — the model of moving focal planes [54-58]. It was established that beyond the point where the beam collapses there is produced not a waveguide regime, but a certain number of foci — regions with very high energy concentration and small dimensions, ~5-10 wavelengths. As a result of the nonstationary character of the laser beams in time, the focal regions move along the axis of the light beam. What were previously assumed to be self-focusing "filaments" are the trails of the "traveling foci."

In the presence of self-focusing in the medium, the conditions for the excitation of SRS are significantly changed. These two phenomena are the subject of a large number of experimental and theoretical studies [2, 25, 59-76].

From the point of view of the angular distribution of SRS in media with self-focusing, great interest attaches to the papers by Garmire [24, 25], who pointed out the existence of rings of class II (see Section 1).

According to Garmire, there are two distinguishing features of class-II radiation.

1. Class-II radiation does not satisfy the phase-synchronism conditions (1.1) and (1.2). To compare the calculated and experimental values of the angles between the axes and the generators of the cones of classes I and II, exact measurements were performed in [24, 25] of the dispersion of a number of investigated liquids. For the cones of class I, satisfactory agreement was obtained between the calculated and experimental data. The angles of the cones of class II in the anti-Stokes region are larger, and in the Stokes region are smaller, than the corresponding phase-synchronism angles, the difference reaching as high as 30%.

2. Class-II rings are much more intense than class-I rings, and are more diffuse. Therefore in those media in which scattering of class II is excited, the class-II rings usually mask the class-I rings. It is important that the intensity of the class-II rings of the first anti-Stokes component is much higher than the intensity of the radiation of the first Stokes component propagating at the synchronism angle. In addition, the anti-Stokes radiation of class II does not depend on the intensity of the Stokes radiation propagating at the phase-synchronism angle to the axis. The rings of the anti-Stokes radiation remain symmetrical and homogeneous when the cell is tilted away from the axis (see Section 1). All this indicates that the class-II anti-Stokes radiation cannot be due to Stokes radiation propagating at the phase-synchronism angle. According to Garmire, this radiation can be excited only by Stokes radiation propagating near the axis of the exciting radiation (at an angle not exceeding 0.01 rad).

3. When the SRS is excited by focused laser radiation, the diameter of the rings of class I varies with the focal length of the focusing length, whereas the diameter of the class-II rings does not depend on this quantity. When laser radiation is focused with a cylindrical lens, the class-I rings turn into ellipses, whereas the class-II rings remain circles. Because of this property of the class-II rings, they can be distinguished experimentally from those of class I.

It is noted in [25] that the angular distribution of SRS has a complicated structure in a number of liquids, namely, the rings of class II of higher order usually split into two or more components; in a number of cases one observes in the anti-Stokes radiation, besides the off-axis radiation, also radiation directed along the axis. It is noted also that the diameters of the rings are not quite stable (the uncontrollable change of the diameter can reach 20%).

The change of the character of the angular distribution of the SRS, according to the data of [77], can serve as a criterion for the onset of self-focusing in a medium. In the cited reference they investigated radiation at anti-Stokes frequencies in acetone and cyclohexane and in a mixture of these substances with carbon disulfide. The content of the carbon disulfide in the mixtures was ~5-10%. In pure acetone, only radiation of class I was observed. In the mixture of acetone with carbon disulfide the character of the angular distribution of the first anti-Stokes component changed. Near the threshold, a sharp "surface radiation" ring was observed with a diameter given by the equation

$$k_1 \cos\theta = 2k_0 - k_{-1}. \tag{2.4}$$

(The theory of surface radiation is considered in [78].) With increasing pump a broad ring of smaller diameter, belonging to class-II radiation, appeared simultaneously with the sharp surface-radiation ring.

Similar observations were made when SRS was excited in cyclohexane and, respectively, in a mixture of cyclohexane with carbon disulfide.

The conclusions drawn in [25, 77] that self-focusing influences the structure of the class-II rings seem too categorical to us. The point is that in the cited studies the SRS was excited in a number of cases by a laser in the multimode regime, and this leads (according to our data) to a complication of the angular distribution of the SRS. We note that in nitrobenzene, in which the self-focusing is very clearly pronounced, the rings of class II remain sharp, a fact that likewise does not agree with the conclusions of the cited papers.

The assumption that the class-II rings are due to emission from self-focusing filaments, advanced in [25, 77], is also ambiguous. For example, in [25] results are presented of experiments in which in pure acetone and cyclohexane they observed sharp rings of class II and class I simultaneously. No special investigations of self-focusing in these liquids were made in [25].

There is no doubt that in many substances self-focusing sets in ahead of SRS. For example, in such substances typically used in SRS experiments as benzene, nitrobenzene, and carbon disulfide, the self-focusing threshold estimate from Eq. (2.2) is, in agreement with experiments [72], 10-20 times lower than the SRS threshold. This means that the SRS phenomenon in these substances takes place in a medium in which the exciting radiation has already been subjected to self-focusing. Thus, the change of views on the nature of the self-focusing makes the assumption that the rings of class II are connected with self-focusing filaments unfounded. Subsequent studies lead to an even more complicated character of the connection between SRS phenomena and self-focusing.

The connection between the distribution of the SRS radiation over the beam cross section (in the "near field") and the phenomenon of self-focusing was investigated in [66, 73-76]. The substances investigated had both large Kerr constants (carbon disulfide) and small Kerr constants (liquid nitrogen, calcite). Self-focusing at the first Stokes component of SRS was observed in liquid nitrogen in the absence of self-focusing at the exciting radiation (SRS self-focusing). The discovery of the SRS self-focusing phenomenon shows that even in substances with small Kerr constants the SRS phenomenon is subject to such complicating nonlinear effects as self-focusing.

There is no doubt that self-focusing and SRS very frequently accompany each other. Self-focusing leads to a redistribution of radiation density and to an increase of the density at individual points of the exciting beam, and this contributes to the appearance of SRS.

A detailed analysis of the entire aggregate of questions connected with self-focusing and excitation of SRS is beyond the scope of the present article. Our task is to trace the relation between these phenomena only for those cases when they have a direct bearing on our experiments.

3. Intensity Distribution in the SRS Spectrum

Usually only a small number of the frequencies from the total Raman scattering spectrum of a given substance appears in the SRS spectra. Most frequently the SRS spectrum contains one of the vibrational frequencies. In some substances, for example in styrene [79, 80], two frequencies are simultaneously excited. In mixtures, likewise at high exciting-radiation powers, frequencies of two components appear [81, 82].

The most essential feature of the SRS spectra is the considerable intensity of the anti-Stokes and Stokes harmonics compared with the harmonics of the spontaneous Raman scattering.

The ratio of the intensities of the various SRS components depends on the experimental conditions. It must be borne in mind that a significant role is played in SRS, besides the radiation processes, also by processes of absorption at various frequencies (inverse Raman scattering). If we recognize that at each Stokes and anti-Stokes frequency the radiation can be produced by different mechanisms and also have different angular characteristics (rings of classes I and II, radiation along the axis of the exciting light), then the overall picture of the distribution of the SRS intensity turns out to be most complicated. An investigation of the distribution of the intensity in the SRS spectra under various excitation conditions casts light on the roles of the different excitation mechanisms. Such investigations, however, are still quite few and insufficiently detailed.

Sokolovskaya et al. [66] investigated the relative intensity of the first and second Stokes lines and the first anti-Stokes line in the CS_2 spectrum at temperatures from +20 to −100°C. The intensities of the lines of the ordinary Raman spectra of CS_2 increase with decreasing temperature [83]. Accordingly the line intensity in the SRS spectrum also increases with decreasing temperature, and becomes radically redistributed among the various components. In particular, they observed in [66] a considerable increase of the relative intensity of the second Stokes component.

The intensities of various SRS components were investigated [84, 85] as functions of the energy of the exciting radiation, using a photographic method for CS_2 (656 cm⁻¹ line) and a calorimetric method for liquid nitrogen (2330 cm⁻¹ line). At low values of the exciting-radiation energy, the intensities of the SRS components increase approximately exponentially, with the argument of the exponential decreasing with increasing number of the harmonic. With increasing E, the rate of growth of the energy of the SRS components decreases. At large values of E, the curves corresponding to the SRS components of lower order reveal a pronounced "saturation" (horizontal section of the curves). Attention is called also to the fact that the energy of the exciting radiation passing through the sample remains practically constant when the energy E of the exciting radiation entering the sample is changed significantly.

In some cases the task is to obtain maximum conversion of the energy of the exciting radiation in a single (first Stokes) SRS component. To solve the problem, a system consisting of a driver laser and amplifier was used in [86]. In such a system, when the propagation directions of the exciting radiation and of the first Stokes components in the amplifier coincided, three regimes were observed: a) linear amplification, wherein the signal pulse retained its waveform; b) the start of saturation, when the amplified pulse deviated from the input pulse but still did not duplicate the pump; c) strong saturation, wherein the output signal duplicated the waveform of the pump pulse. The signal quantum yield at the first Stokes frequency then approached 100%.

From the point of view of identifying the principal SRS mechanism, a great interest attaches to data on the correlation between the intensity of the first anti-Stokes component and the first Stokes component [84].

The scattered radiation was registered at angles 20-25°, i.e., the intensity integrated over the angles was registered. The intensity of the first anti-Stokes component increased rapidly with increasing intensity of the first Stokes component. This function agrees well with the mechanism of the Stokes − anti-Stokes four-photon coherent process, which will be considered in Chapter IV.

The mechanism of successive excitation of various SRS components was treated in [87] as a parametric process. The SRS at the Stokes component is in this case a parametric process due to interaction of the light waves with the phonon waves, while at the anti-Stokes component it constitutes parametric interaction of four waves: laser, Stokes, anti-Stokes, and phonon. It is assumed that the laser wave is plane and that the power of the laser radiation does not vary over the length of the cell. Under these assumptions they solved in [87] a system of equations that connect the intensities of Stokes components of different orders for different beam path lengths in the medium. Numerical computer calculations were performed for nitrobenzene. It was shown that as the radiation propagates along the cell the energy is being transferred to higher and higher Stokes components. Regions in which the waves of only one component exist alternate along the cell with overlap regions containing two components, with the intensity of one of the waves decreasing rapidly, and that of the other increasing.

The theory developed in [88] agrees qualitatively with experiment. However, such a distinguishing feature of SRS as the almost constant ratio of the intensities of the exciting radiation, of the first Stokes component, and of the second Stokes component contradicts the conclusions of this theory.

The initial premises of this theory were refined in [89] to conform with the real conditions of the experiment. In particular, account was taken of the fact that some of the intensity of the exciting radiation becomes distributed over the cross section of the cell (radial distribution). Computer calculations were performed for carbon disulfide under the assumption that the radial distribution is Gaussian. The calculations have shown that all the Stokes components are simultaneously present at the exit from the cell (if the exciting radiation is strong enough).

Experiment [88] showed good agreement with the calculations by the theory of [87]. The object of investigation in [88] was carbon disulfide, which is a self-focusing liquid. To exclude the influence of self-focusing and of stimulated Mandelstam—Brillouin scattering (SMBS) on the SRS process, the SRS was excited by short pulses of duration τ from 0.6 to 1.5 nsec, not long enough for the indicated accompanying processes to develop.

The efficiency of conversion of the radiation into the various SRS components reached 60%. With increasing τ, this efficiency decreased and the intensity of the SMBS increased.

In the papers cited above, no account was taken of four-photon processes of SRS excitation. Without going into details, we mention that the theory of these processes was developed in [7, 90-92].

As indicated above, to ascertain the real mechanisms that lead to the onset of different SRS components, great interest attaches to the ratio of the intensity of the radiation propagating in the direction of the exciting radiation into the cones.

An attempt to explain the singularities of the anti-Stokes radiation propagating along the axis (i.e., in the direction of the exciting light) and into the cones was made in [31, 32, 93].

In [93] they investigated carbon disulfide, benzene, cyclohexane, acetone, as well as a mixture of acetone with carbon disulfide. In carbon disulfide, the picture of the angular distribution always had a smeared out and diffuse character. A uniformly distributed background was observed inside a blurred ring. The central spot was only lightly discerned against this background. Clear-cut photographs were obtained for benzene. For the first anti-Stokes component, the intensity of the radiation propagating along the axis was approximately equal to the intensity of the radiation propagating in the cones. Axial radiation was observed also in the distribution of the second anti-Stokes component, but its intensity was lower than for the radiation going into the cones.

In cyclohexane, at low values of the pump energy, the radiation produced was that of the first anti-Stokes component and propagated along the axis. With increasing pump energy, anti-Stokes radiation was produced and propagated in a cone whose intensity increased rapidly with increasing pump energy. Similar results were obtained for acetone.

Morozova [32] investigated in detail the anti-Stokes radiation propagating along the axis and in the cones, in substances with small Kerr constants, namely liquid nitrogen and calcite. It was observed that at low pump energies the first anti-Stokes component is produced and is directed only along the axis. With increasing pump energy, an anti-Stokes ring also appears. Further increase of the pump leads to an increase of the intensity of the ring and to a decrease of the intensity of the central spot, which vanishes at still pump energies.

A similar change in the ratio of the intensities of the central spot and of the ring results from a decrease in the focal length of the lens that gathers the laser radiation into the scattering medium. At a large focal length of the lens, only a central spot is observed. Decreasing the focal length leads first to the appearance of a ring simultaneously with the central spot. At short focal lengths, only the ring is observed.

Morozova [32] investigated also the influence of the length of the cell with the liquid nitrogen on the angular distribution of the SRS. It turns out that an increase of the length of the cell causes the same change in the picture of the angular distribution as an increase in the energy of the exciting radiation, but the changes of this picture manifest themselves less strongly than when the energy is changed.

The results of [32] are attributed to the presence of a unique SRS self-focusing in media with small Kerr constants, with the self-focusing assuming an increasing role when the energy density of the exciting radiation is decreased.

The angular distribution of the intensity of the Stokes components of SRS in benzene and in carbon disulfide was investigated in [94-96]. The SRS was excited by a harmonic of a YAG-Nd[3] laser at a wavelength 0.53 μm. Owing to partial mode locking, the exciting radiation comprised a sequence of short pulses of 0.5 nsec duration. At such a short exciting-pulse duration, the SRS in the benzene was not encumbered by accompanying SMBS and self-focusing (it was impossible to get rid of the self-focusing in CS_2).

In benzene, at a cell length 20 cm, the second Stokes component was produced in the form of a beam having the same directivity and the same divergence as the first Stokes component (4 mrad). The threshold value of the pump at the wavelength 5300 Å was 150 MW/cm^2 for this process. At a cell length 10 cm, the SRS threshold increased to 300 MW/cm^2 and the second Stokes component was radiated into a cone whose generator made an angle $2.4 \pm 0.2°$, or 0.040 ± 0.003 rad, with the axis. This angle corresponds to off-axis radiation of class II. Thus, class-II radiation was observed in the cited studies in the absence of self-focusing, a fact which we consider to be worthy of much attention. The vertex of the cone was located at the end of the cell. When the pump power was increased by 1.3 times, radiation was produced both in a cone and along the axis. The authors attribute the difference they observed in the angular distribution of SRS to the different geometrical conditions of the excitation of the SRS in the long and short cells.

Judging from the published data cited in the present section, we can conclude that the laws governing the formation of axial SRS radiation and radiation into cones are different.

4. Theory of the Angular Distribution of the SRS Components

As follows from the review of the experimental material, the laws governing the angular distribution of SRS are quite complicated and varied. The emission of each SRS component follows some selected direction, and a strong concentration of the energy of the scattered radiation along this direction is observed. The ratio of the intensities of the radiation propagating along the axis of the exciting radiation (central spot) and at different angles to the axis (rings of various classes) depends on the singularities of the nonlinear medium and on the excitation conditions.

The task of the theory of the angular distribution of SRS is to explain the predominant concentration of the scattered radiation along certain selected directions and the laws governing the partition of the intensities among the axial radiation and the radiations in the cones of various classes. It should be noted that only the first of these problems has been dealt with in detail up to now.

1. Semiclassical Theory of SRS. Some features of the angular distribution of SRS can be understood even on the basis of a very simple semiclassical treatment of the process of SRS production [20, 21, 97].

We consider a molecule with polarizability α, situated in an electric field $\mathbf{E}$. Under the influence of this field, the molecule acquires an electric moment $\mu = \alpha \mathbf{E}$ and a potential energy $u = \frac{1}{2}\alpha E^2$.

Let x be the vibrational coordinate describing a certain vibrational process in the molecule. Assuming that the molecule is acted upon by a driving force of

$$F = -\frac{\partial u}{\partial x} = \frac{1}{2}\frac{\partial \alpha}{\partial x}E^2, \tag{4.1}$$

we obtain the equation of the intramolecular vibrations

$$m\ddot{x} + R_0\dot{x} + fx = F = \frac{1}{2}\frac{\partial \alpha}{\partial x}E^2 \tag{4.2}$$

or

$$m\ddot{x} + R_0\dot{x} + fx = F_0\cos\omega t \tag{4.3}$$

(the driving force is assumed harmonic). The solution of this equation for the frequency ω close to the resonant frequency $\omega_r = \sqrt{f/m}$ is

$$x = \frac{F_0}{R_0\omega}\sin\omega t. \tag{4.4}$$

Here R_0 is a phenomenological damping constant; f is the quasielastic constant of the considered molecule, and corresponds to a normal vibration with frequency ω_r.

We assume now that the electric field $\mathbf{E}$ is an aggregate of a certain number of plane waves that differ in frequency by an amount ω or (in the more general case) by a multiple of ω. In the simplest case of two such waves

$$\mathbf{E} = \mathbf{E}_0 e^{i(\omega_0 t - \mathbf{k}_0 \mathbf{r})} + \mathbf{E}' e^{i(\omega' t - \mathbf{k}' \mathbf{r} + \varphi')}, \tag{4.5}$$

where $\omega_0 - \omega' = \omega$. In this case

$$E^2 = E_0^2 + E'^2 + 2E_0 E'\cos[(\omega_0 - \omega')t - (\mathbf{k}_0 - \mathbf{k}')\mathbf{r} - \varphi']; \tag{4.6}$$

the constant terms can be left out here, and we obtain

$$F_0 = \frac{d\alpha}{dx}\mathbf{E}_0\mathbf{E}'. \tag{4.7}$$

Accordingly

$$x = \frac{\mathbf{E}_0\mathbf{E}'\frac{d\alpha}{dx}}{R_0(\omega_0 - \omega')}\sin[(\omega_0 - \omega')t - (\mathbf{k}_0 - \mathbf{k}')\mathbf{r} - \varphi']. \tag{4.8}$$

10

The molecular vibrations indicated above produce an oscillating dipole moment

$$\mu = x \frac{d\alpha}{dx} \mathbf{E} = \frac{E_0 E' \left(\frac{d\alpha}{dx}\right)^2}{R_0 (\omega_0 - \omega')} \sin\left[(\omega_0 - \omega')t - (\mathbf{k}_0 - \mathbf{k}')\mathbf{r} - \varphi'\right] \mathbf{E}. \tag{4.9}$$

The rate of energy exchange between the dipole moment and the field component with frequency ω' is given by

$$P' = -\left\langle \frac{d\mu}{dt} \mathbf{E}' \right\rangle, \tag{4.10}$$

where the averaging is over the time. From this we get the power transferred to the component $\mathbf{E}'$ from the initial field $\mathbf{E}_0$:

$$P' = \frac{1}{2R_0} \left(\frac{d\alpha}{dx}\right)^2 \frac{\omega'}{\omega_0 - \omega'} (E_0 E')^2. \tag{4.11}$$

For Stokes radiation $\omega' = \omega_0 - \omega_2$, $P' > 0$, and the radiation corresponding to the components $\mathbf{E}'$ is enhanced. For anti-Stokes radiation $\omega' = \omega_0 + \omega_2$ and this radiation loses energy.

To explain the possibility of generation of anti-Stokes lines, we consider the case when the field $\mathbf{E}$ has three components with different frequencies:

$$\mathbf{E} = \mathbf{E}_0 e^{i(\omega_0 t - \mathbf{k}_0 \mathbf{r})} + \mathbf{E}_{-1} e^{i[(\omega_0 - \omega_2)t - \mathbf{k}_{-1}\mathbf{r} + \varphi_{-1}]} + \mathbf{E}_1 e^{i[(\omega_0 + \omega_2)t - \mathbf{k}_1 \mathbf{r} + \varphi_1]}. \tag{4.12}$$

By the same method as above we calculate the vibrations of the molecule and its oscillating dipole moment. The enhancement of the power of the Stokes and anti-Stokes radiation is given by the equations

$$P_{-1} = \frac{1}{2R_0} \left(\frac{d\alpha}{dx}\right)^2 \frac{\omega_0 - \omega_r}{\omega_r} \{(E_0 E_{-1})^2 + (E_0 E_1)(E_0 E_{-1}) \cos\left[(2\mathbf{k}_0 - \mathbf{k}_1 - \mathbf{k}_{-1})\mathbf{r} + \varphi_1 + \varphi_{-1}\right]\}, \tag{4.13}$$

$$P_1 = \frac{1}{2R_0} \left(\frac{d\alpha}{dx}\right)^2 \frac{\omega_0 + \omega_r}{\omega_r} \{- (E_0 E_1)^2 - (E_0 E_1)(E_0 E_{-1}) \cos\left[(2\mathbf{k}_0 - \mathbf{k}_1 - \mathbf{k}_{-1})\mathbf{r} + \varphi_1 + \varphi_{-1}\right]\}. \tag{4.14}$$

Thus, if $|\mathbf{E}_{-1}| > |\mathbf{E}_1|$ then the component $\mathbf{E}_1$ can be amplified if the following conditions are satisfied:

$$2\mathbf{k}_0 = \mathbf{k}_1 + \mathbf{k}_{-1}, \qquad \cos(\varphi_1 + \varphi_2) < 0. \tag{4.15}$$

Here $\mathbf{k}_0$, $\mathbf{k}_{-1}$, and $\mathbf{k}_1$ are the wave vectors of the initial, Stokes, and anti-Stokes waves, respectively. If $\varphi_1 + \varphi_{-1} = \pi$, then the enhancement of the anti-Stokes lines is largest. The amplification of the Stokes line decreases for the corresponding direction because of the negative sign of the cosine.

In the presence of Stokes radiation of frequency $\omega_0 - \omega_r$, a field $\mathbf{E}_{-2}$ with frequency $\omega_0 - 2\omega_r$ can be produced, and so on.

A shortcoming of the approach described above is that emission of SRS components is possible in a wide angle interval. Experiment, however, shows these components to be emitted in a narrow angle interval, near directions determined by the phase-synchronism conditions.

We note that in [20, 21, 97] the SRS was analyzed on the basis of an investigation on the interaction of plane waves present in the medium; the sources exciting these waves were not considered. Real sources of waves with frequency that differs from the exciting frequency are molecules of scattering matter (priming sources). In the case of SRS, the light wave incident on the medium is so intense that it alters substantially the electrodynamic properties of the medium. Therefore the radiation of the scattering molecules must be considered in a medium with altered characteristics. This approach to the SRS theory was developed in a number of papers by Lugovoi [8, 35, 98-101]. In Lugovoi's theory, on the one hand, account is taken of the real nature of the priming sources of the Stokes-frequency radiation, and on the other hand, of the driving part of the effect (via negative absorption in the medium, which enters in the material equation of the medium). The total SRS intensity is obtained by summing the radiation intensities from all the priming sources.

Thus, in Lugovoi's theory only the pump is a plane wave, while the scattered radiation is generated by individual molecules—dipoles. Each of them emits at the first Stokes frequency.

The intensity, angular distribution, and other characteristics of the SRS were determined in [8, 35, 98-101]. Both the first anti-Stokes component and all the higher Stokes and anti-Stokes components of the SRS are

11

amplified in the medium together with the first Stokes component. According to Lugovoi's theory, the gains of all the components are determined by the gain of the first Stokes component.

The emission angles of the SRS components and the corresponding angles of absorption of the first Stokes components are determined in accordance with the Lugovoi theory by the relations

$$\mathbf{k}_m + m\mathbf{k}_{-1}^{(m)} = (m+1)\mathbf{k}_0,$$
$$\mathbf{k}_{-m} + (m-1)\mathbf{k}_0 = m\mathbf{k}_{-1}^{(-m)}. \tag{4.16}$$

It is important that, in contrast to Townes's theory [20, 21], Eqs. (4.16) determine the scattering of some component of the SRS in a narrow angle interval. In this sense the Lugovoi theory agrees better with experiment than the elementary theory developed at the beginning of this section. We note that class-II radiation is not considered in either case.

Lugovoi's theory leads also to some conclusion concerning the relative intensity of the various SRS components. For example, it is indicated in [8] that the maximum energy density of the first anti-Stokes component must be of the same order of magnitude as the energy density of the first Stokes component.

To explain the complex angular distribution of the intensity of the anti-Stokes SRS components, which leads to formation of the rings of first and second class, Shimoda [102, 103] has considered the scattering in filamentary regions produced in the medium under the influence of high-power laser radiation.

The theoretical values of the angles of the class-II cones outside the liquid are compared with the experimental values in Table 1.

It can be seen that the experimental and theoretical data are in good agreement.

As indicated in Section 2, our understanding of the self-focusing phenomenon has changed significantly of late. According to the theory of Lugovoi and Prokhorov [54-57], which is corroborated by numerous experiments, self-focusing produces in the medium traveling foci whose trails were taken to be previously the self-focusing "filaments." Therefore the initial premises of Shimoda's theory cannot be regarded as incontrovertible, and the theory as a whole requires a critical approach. We note that the good agreement between the angles of the class-II radiation cones calculated by Shimoda and the experiment cannot serve as a decisive argument favoring this theory, since it makes use of a parameter whose value is obtained from data on the angular distribution. Nonetheless, the Shimoda theory is of undoubted interest as the first qualitative attempt of interpreting the class-II SRS radiation.

2. Quantum Theory of SRS. The semiclassical theory of the SRS phenomenon, which was described above, gives a certain idea of the mechanism of the onset of the harmonics and of the anti-Stokes lines. This theory, however, gives no indication of a quantitative determination of the intensity in the SRS spectra. The more rigorous quantum theory cannot yet be regarded as complete, despite the presence of a number of papers on the subject (see, e.g., [7, 90]). We note that the existing papers on the quantum theory of SRS hardly touch upon the angular distribution of the SRS radiation. We present here therefore only some estimates of the SRS threshold in accordance with [7]; these will be needed later on. We consider the two SRS mechanisms that seem to be the most probable.

a. Stepwise excitation of harmonics. The primary act of Stokes Raman scattering of light results in an excited molecule and in a photon with decreased frequency $\hbar\omega_{-1}$. The Stokes interaction of the photon with the unexcited molecule yields a photon $\hbar\omega_{-2}$, where $\omega_{-2} = \omega - 2(\omega - \omega_{-1})$, i.e., the first harmonic is produced. These processes are repeated and produce harmonics of ever higher order in the Stokes region. The interaction of the $\hbar\omega$ exciting-light photon with the excited molecule produces an anti-Stokes photon $\hbar\omega_1$, and the molecule goes over into the unexcited state. The interaction of the anti-Stokes photon $\hbar\omega_1$ with the excited molecules produces an anti-Stokes harmonic with frequency $\hbar\omega_2$, where $\omega_2 = \omega + 2(\omega - \omega_{-1})$, etc. The competing processes are the scattering of the anti-Stokes photons by the excited molecules with formation of photons with decreased frequency, and the scattering of the Stokes photons by the excited molecules, as a result of which photons with increased frequency are produced.

b. Four-photon SRS excitation processes. According to the four-photon SRS-process scheme proposed by Terhune [15] in this process two photons are absorbed and two new photons are emitted. In the simplest case two photons of the exciting radiation are absorbed and photons of the anti-Stokes and Stokes components are emitted:

$$2\mathbf{k}_0 \rightarrow \mathbf{k}_{-1} + \mathbf{k}_1. \tag{4.17}$$

TABLE 1. Class-II Radiation in Benzene (angles in units of 10^{-2} rad) according to the Data of Shimoda [102]

Order	First cone		Second cone		Order	First cone		Second cone	
	theory	experiment	theory	experiment		theory	experiment	theory	experiment
−3	5.63	—	3.78	—	1	3.24	3.25	—	—
−2	3.64	3.8	—	—	2	5.93	6.00	4.85	4.8
−1	—	—	—	—	3	8.44	8.5	7.74	7.6

The four-photon SRS processes can be either coherent (with certain synchronism conditions satisfied) or noncoherent (without satisfaction of these conditions). We consider below noncoherent four-photon SRS processes. These processes can take place either without or with a change of the vibrational state of the molecule.

In accordance with the Raman-scattering theory developed in [7], the four-photon process that leaves the molecule in the initial state can proceed in the form of two independent but simultaneous processes of Stokes and anti-Stokes Raman scattering. The probability W_{4ph} of such a complicated process is equal to the product of the probabilities of the Stokes and anti-Stokes scattering:

$$W_{4ph} = W_{St} W_{aSt}. \tag{4.18}$$

It is easily seen therefore that the probability of the four-photon process is generally speaking much lower than the probabilities of the two-photon scattering processes.

If the molecule goes over in the four-photon process into an excited vibrational state L, then we have for the probability of this process, in accordance with [7],

$$W_{cr}^{I} = \frac{|H_{1L}'|^2 |H_{Lk}'|^2 |f_{k1}|^2}{|E_1' - E_L'|^2 |E_k' - E_1'|^2}. \tag{4.19}$$

The matrix element H_{1L}' can be calculated as a "composite" matrix element of the $1-L$ transition, accompanied by Raman scattering, with transition through the excited electronic states of the molecule l (1 and k are the initial and final states of the molecule).

We now compare the various considered processes of excitation of SRS harmonics. It is convenient to use for comparison the value of the threshold π of the excitation of a given component (with π_1 denoting the threshold of the excitation of the first Stokes component of the SRS for the 992 cm^{-1} benzene line). If we introduce the notation $a = Kl$, where l is the length of the cell and K is the gain, then we have approximately

$$\pi_1/\pi_2 = a/a_1. \tag{4.20}$$

For benzene, according to [7], we can assume a tentative value $a_1\pi_1 = 12$.

In stepwise excitation of the second Stokes component in benzene we have, according to estimates given in [7],

$$(a\pi)_{-2} \approx 3a_1\pi_1. \tag{4.21}$$

This value agrees satisfactorily with the available experimental data. For the first anti-Stokes component, in successive excitation, it can be assumed that the scattering is by molecules in the excited vibrational state. If we assume for the concentration C_1^* of the molecules in the excited vibrational state its equilibrium value, i.e. (for the 992 cm^{-1} of benzene), $C_1^* = 0.008$, then we get for the threshold of the first anti-Stokes component in benzene $(a\pi)_{step}^2 \approx 120a_1\pi_1$. The concentration C_1^* increases in the SRS process, but it is difficult to assume that it changes substantially. Thus, the considered process can play a significant role only at very high exciting-radiation powers.

We consider now the four-photon SRS excitation processes. For the process without change in the vibrational state of the molecule, if the radiation propagates through a channel having a cross-sectional area $S = 10^{-5}$ cm^2, the necessary power of the exciting radiation is

$$P_{exc} = 1.5 \text{ MW}.$$

However, the process proceeds at an exciting-radiation power only under the condition that a Stokes radiation of power $P_{-1} \approx 1$ MW, obtained by some other means, is already present at the synchronism angle. An autonomous four-photon noncoherent process has a practically unattainable threshold

$$P_{exc} = 10^{15} \text{ MW}.$$

13

An estimate of the four-photon SRS processes in which the vibrational state of the molecule changes leads to the conclusion [7] that at $P_{exc} \doteq 10$ MW and $P_{-2} \approx 1$ MW the threshold can be reached only at $S \approx 10^{-7}$ cm^{-2}.

Thus, the considered four-photon processes can play an essential role if thin light-conducting channels with very high energy density are produced in the medium under consideration. The existence of such channels depends on the real geometry of the experiments in which the SRS is excited. This group of questions is the subject of an exhaustive paper [104]. It appears that channels of diameter 0.1-0.01 mm do exist in those experiments in which the exciting radiation is focused into the cell with short-focus lenses. The real role of self-focusing in the processes of excitation of SRS cannot be regarded as completely clear at the present time, all the more since in addition to the self-focusing of the exciting radiation an important role can be assumed also by SRS self-focusing, which was discovered by Sokolovskaya and co-workers [75, 76].

The noncoherent four-photon processes considered above cannot explain the complicated angular distribution of the SRS radiation. As shown in [105-110], the observed angular distribution of the Stokes and anti-Stokes components of the SRS can be explained by taking into account the contribution made to the SRS by the four-photon coherent processes.

CHAPTER II

EXPERIMENTAL SETUP AND PROCEDURE

5. Experimental Procedure in the Investigation of the Angular

Distribution of SRS near the Axis of the Exciting Radiation

To investigate the angular distribution of the radiation of the SRS components near the axis of the exciting radiation and to investigate the off-axis radiation, an experimental setup was constructed which made it possible to excite SRS in various organic liquids and in calcite crystals.

The light source for the excitation of the SRS was a single-pulse ruby laser. The ruby crystals were 120 mm long and 8 to 12 mm in diameter. The ruby crystal was pumped by two xenon flash lamps of type IFP-2000, which were placed in a double-ellipse illuminator. The reflecting surface of the illuminator was polished and silvered. The ruby crystal was placed in the common focus of the two ellipses of the reflector. The illuminator was cooled with running water and was mounted on a special adjusting stage that was well insulated from the ground to ensure ignition of the lamps by an external pulse. The pump lamps were connected in series. They were fed from a capacitor bank with total capacitance 1500 μF. The capacitor bank was charged from a high-voltage rectifier.

The capacitor bank was discharged through the lamps by applying a pulse from the ignition block. The laser cavity was made up of a flat mirror with a multilayer dielectric coating having a reflection coefficient $R = 99.5\%$ at a wavelength 6943 Å, and one or two plane-parallel quartz plates 10 mm thick. The cavity length ranged from 40 to 60 cm.

To investigate the influence of the various laser operating conditions on the angular distribution of the first Stokes component of the SRS, the investigations were made with two different setups (setups I and II).

A block diagram of setup I is shown in Fig. 1. This setup used a laser Q-switched with a solution of cryptocyanine in ethyl alcohol, with a transmission coefficient $\approx 30\%$ at a wavelength 6943 Å. The solution was contained in a plane-parallel cell 4 of thickness 10 mm, placed between the mirror with $R = 99.5\%$ and the end face of ruby 5.

The spectral composition of the laser emission was investigated by means of interferograms obtained with the Fabry–Perot etalon 12 with an air-layer thickness 3 or 7 cm. The interferograms were photographed in the focal plane of UF-82 camera 13. The width of the spectral line was $\Delta\nu = 0.015$ cm^{-1}. It should be noted that the use of saturable filters leads to narrowing of the lasing spectrum. By using additional mode selections with the aid of passive interferometers in the form of plane-parallel plate 7, which serve as the exit mirror of the cavity, generation of a single longitudinal mode was attained.

The selection of the transverse oscillation modes was effected by diaphragm 6 of 1-1.2 mm diameter, placed between the end face of the ruby and the quartz plates.

The duration and waveform of the exciting-radiation pulse was monitored by a system consisting of an FÉK-09 coaxial photocell 2 and an I2-7 nanosecond time-interval meter 1. The pulses were photographed from

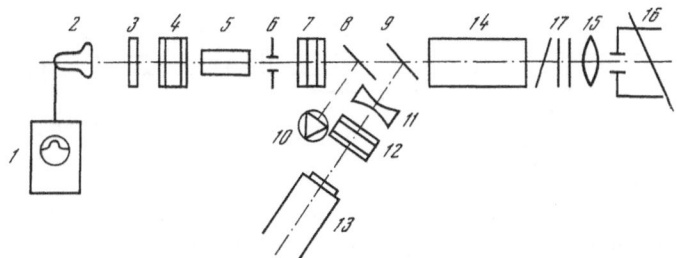

Fig. 1. Diagram of the setup for SRS excitation in the single-mode regime. (1) I2-7 oscilloscope; (2) FÉK-09 coaxial photocell; (3) dielectric mirror (~99.5% reflection at $\lambda = 6943$ Å); (4) cell with solution of cryptocyanine in ethyl alcohol; (5) ruby rod; (6) diaphragm; (7) two plane-parallel quartz plates; (8, 9) turning plates; (10) calorimeter; (11) scattering lens; (12) Fabry—Perot etalon; (13) camera with f = 800 mm; (14) cell with substance or crystal; (15) lens with f = 120 mm; (16) spectrograph; (17) set of filters to attenuate the SRS.

the I2-7 screen on RF-3 photographic film. The pulse duration at half-intensity was 20 nsec. The waveform of the pulse was very close to Gaussian. The laser energy was monitored with calorimeter 10 graduated against an IMO-1 standard power meter. The recording instrument was an M197/2 galvanometer with sensitivity $(2.16 \pm 0.2) \cdot 10^{-3}$ J/division.

The maximum radiation energy was on the order of 0.4 J (in the case when one longitudinal mode was generated) and 0.03 J (when one transverse mode was generated).

The laser cavity was adjusted with an autocollimator or with a helium—neon laser. The mirror with transmission coefficient 99.5%, the plane-parallel plates, and the end faces of the ruby crystal were mounted parallel to one another. Ruby rods with end faces tilted 1.5-2° were sometimes used.

The energy, waveform, and duration of the pulse, as well as the laser emission spectrum, were monitored simultaneously. The divergence of the laser beam was measured in the focal plane of a lens of focal length f = 1 m. The photographic-film density was converted into intensities with the aid of markers photographed on the particular type of film. In the case when one transverse mode was generated the diameter of the beam leaving the resonator was $\simeq 1.2$ mm and the beam divergence, equal to $8.1 \cdot 10^{-4}$ rad, was close to the divergence determined by the diffraction limit.

In investigations of the angular distribution of the first Stokes component of benzene ($\Delta\nu = 992$ cm^{-1}), carbon disulfide ($\Delta\nu = 656$ cm^{-1}), and acetone ($\Delta\nu = 2921$ cm^{-1}) the SRS spectra were excited by an unfocused beam in the cell 14. The cells ranged in length from 10 to 40 cm. The cell windows were inclined 5-10° to the axis. This excluded the possibility of generation of SRS by light reflected from the cell windows. There was no feedback from the cell to the laser, as was verified by experiment. The cell was placed 4 m away from the laser. The SRS radiation was focused by lens 15 on the plane of the slit of spectrograph 16. During the course of operation the slit was removed. The spectrograph had a collimator with focal length 180 mm, a camera with focal length 270 mm, and a diffraction grating with 600 lines/mm as the dispersing element.

The angular distribution of the intensity of the first Stokes SRS component was measured by photometry of the obtained photographs.

The angle divergence of the SRS radiation and of the exciting radiation at the exit from the cell were determined from the half-width of the observed distribution d, using the formula

$$\psi = \frac{1}{K} \frac{d}{f},$$

where f is the focal length of the lens that gathers the light in the plane of the slit of the spectrograph; K is the magnification of the spectrograph.

To assess the influence of multimode laser emission on the angular divergence of the emission of the first Stokes component, we used setup II. A block diagram of the setup is shown in Fig. 2.

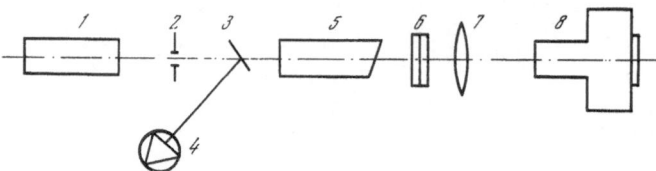

Fig. 2. Diagram of setup for the excitation of SRS in the multimode regime. (1) Laser; (2) diaphragm; (3) glass plate; (4) calorimeter; (5) cell with investigated liquid; (6) set of color and neutral filters; (7) lens with f = 80 mm; (8) prism spectrograph without entrance slit.

In this setup we used a ruby laser in which the Q switching was by means of a rotating total-internal-reflection prism. The pulse duration at half-intensity was 30-35 nsec. The maximum radiation power was 30 MW. No mode selection was effected. The beam diameter at the laser output exceeded 12 mm, so that we were able to illuminate diaphragms 2 up to this height. The spatial distribution of the beam intensity was inhomogeneous. We investigated the influence of the inhomogeneities of the intensity on the angular characteristics of the SRS.

6. Experimental Procedure in the Study of Off-Axis Class-II SRS Emission

Class-II emission in organic liquids (benzene, carbon disulfide, and nitrobenzene) was investigated with the setup shown in Fig. 3. The choice of the objects of the investigations was dictated by the fact that their refractive indices are known with high accuracy in a wide spectral range.

The light source for the excitation of the SRS spectra was a ruby laser operating in the single-pulse regime. One axial mode was registered in the laser emission. A detailed description of the laser characteristics is given in Section 5.

Inasmuch as the off-axis emission is produced not only by the exciting radiation but also by the SRS components, we have investigated the influence of the intensities of the first and second Stokes components propagating along the axis of the exciting radiation on the formation of the off-axis emission of the second, third, and fourth Stokes components of the SRS. To carry out these investigations, we included in the setup, besides the principal cell 15, also an auxiliary cell 9 with the same liquid as in cell 15.

The laser emission was first focused by lens 8 into cell 9, 200 mm long, mounted at an angle to the axis. This excludes the possibility of lasing as a result of light reflection from the cell windows (one of the reflections was incident on the blackened lateral face of the cell). The radiation emerging from the cell 9 was gathered by lens 10 and passed through diaphragm 11 of 12 mm diameter. This radiation was further directed by beam-splitting plate 13 to spectrograph 18. The selection of the spectral composition of the radiation emerging from cell 9 was effected by means of a set of filters 12, and the spectrum was registered photographically in spectrograph 18. The radiation emerging from the cell 9 had a complicated spectral composition. For example, in the investigation of SRS in benzene, it consisted of radiation at the ruby-laser wavelength $\lambda = 6943$ Å, of the first Stokes component with $\lambda = 7457$ Å, and of the second Stokes component with $\lambda = 8050$ Å. The radiation at the frequencies of the third and fourth Stokes components and anti-Stokes components was removed by a set of selective filters 12. The radiation at the frequency of the second Stokes component was attenuated 70% with filters, when necessary. The radiation emerging from the cell 9 was focused by lens 14 into cell 15, which was also mounted at an angle to the axis.

The radiation from cell 15 was gathered by lens 16 and focused in the plane of the slit of spectrograph 17, the slit being removed in the experiments. The SRS spectra were registered photographically in the focal plane of the camera of the spectrograph 17 on photographic film or plates.

The setup with two cells, which we used, turned out to be very convenient. It was easy to change in it the spectral composition of the radiation entering the main cell 15. In the presence of radiation at the second Stokes frequency, it was easy to excite axial and off-axis SRS of higher Stokes components (we note that prior to our study these components were never investigated by anyone).

When necessary we were able to get rid of the second Stokes component by strongly attenuating it with suitable filters. This has made it possible to study the influence of the first Stokes component on the formation

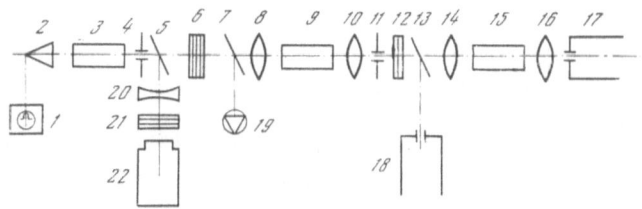

Fig. 3. Diagram of setup for the investigation of off-axis SRS emission. (1) I2-7 oscilloscope; (2) FÉK-09 coaxial photocell; (3) ruby laser; (4, 11) diaphragms; (5, 7, 13) turning plates; (6) set of filters to attenuate the exciting radiation; (8) lens with f = 500 mm; (9, 15) cells with the material; (10) lens with f = 800 mm; (12) set of filters for SRS selection; (14) lens with f = 140 mm; (16) lens with f = 80 mm; (17, 18) spectrographs; (19) calorimeters; (20) diverging lens; (21) Fabry—Perot interferometer; (22) camera with f = 800 mm.

of the off-axis radiation of the second Stokes component of the investigated liquid.

Figure 4 shows the dependence of the intensity of the first Stokes component of the exciting radiation passing through the cell 9 on the energy of the exciting radiation entering this cell. The laser energy was attenuated with a set of calibrated glass filters 6.

It is seen from the obtained curves that the energy of the exciting radiation passing through the cell 9 remains relatively constant when the exciting radiation entering the cell is changed by a factor 3.5, while the intensity of the first Stokes component increases sharply.

This property of the emerging radiation has enabled us to investigate the characteristics of the SRS produced in the second cell, as functions of the intensity of the first Stokes component at a constant intensity of the exciting radiation.

We investigated the dependence of the angles of the cones of the off-axis SRS radiation in benzene, carbon disulfide, and nitrobenzene on the focal length of the lens that focused the exciting radiation into the substance, and on the length of the cell. We used lens with focal lengths 120, 140, 180, 300, 800, and 1100 mm and cells 100, 200, and 300 mm long. We have observed that the emission-cone angles do not depend on the focal length of the lens used to focus the exciting radiation into the medium or on the lengths of the cells. This verifies that the investigated radiation is of class II, since the angles of class-I radiation depend strongly on the focal length of the lens. With the third SRS Stokes component in nitrobenzene as an example, we investigated the influence of the mode composition of the exciting radiation on the structure of the rings. In the case when the laser radiation contained one longitudinal mode and a large set of angular modes, the rings of the class-II radiation had a complicated composition consisting of closely lying thin rings, or else became diffuse. In the case of a single-mode laser, on the other hand, we obtain one or two sharp rings.

We have also investigated the self-focusing of the exciting radiation and of the first SRS Stokes component in the same liquids in which we investigated the angular distribution of the SRS. To this end we investigated the distribution of the intensity in the beam cross section "in the near field." For this purpose, the plane adjacent to the exit window of cell 15 was projected on the plane of the slit of the spectrograph 17 with a magnification 8×. The presence of self-focusing was revealed by the appearance of bright spots into which the bright beam of the exciting radiation broke up. Typical self-focusing points were observed in benzene, carbon disulfide, and nitrobenzene.

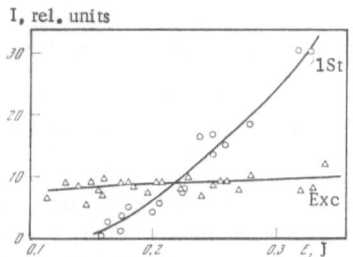

Fig. 4. Dependence of the intensity of the SRS first Stokes component in benzene (curve 1St) and of the exciting radiation passing through cell 9 (curve Exc) on the energy of the exciting radiation.

CHAPTER III

INVESTIGATION OF THE ANGULAR DISTRIBUTION OF THE EMISSION OF
THE SRS COMPONENTS NEAR THE AXIS OF THE EXCITING RADIATION

7. Results of Measurements of the Angular Distribution
of the First SRS Stokes Component

Systematic investigations of the angular distribution of the first SRS Stokes component near the axis of the exciting-radiation beam were initiated by us in 1967, when only sketchy data on this question could be found in the literature. It was noted (see Section 1) that in the case of SRS excitation outside the resonator, the first Stokes component is directed mainly "forward" along the beam of the exciting radiation and has a very narrow angular distribution. There were no quantitative data, however. Our very first studies [105, 106] have shown that to explain the very small angle divergence of the first SRS Stokes component it is necessary to invoke new physical concepts concerning the process of SRS excitation. It was indicated that a role can be played in this phenomenon by coherent four-photon processes. The subsequent development of the theory proceeded in parallel with the experimental research on the angular distribution of the first Stokes component, and later on also on higher SRS components.

In our investigations we did not consider the angular distribution of the radiation when SRS was excited in resonators, or when SRS was generated. These questions are the subject of a number of our other papers [10, 111–115].

We investigated the effect exerted on the angular distribution of the first Stokes component of the SRS by such factors as the mode composition of the excited radiation, the geometrical conditions of the SRS excitation, and the self-focusing of the exciting radiation in the medium [106, 109]. The SRS was investigated in benzene ($\Delta \nu = 992$ cm^{-1}), carbon disulfide ($\Delta \nu = 656$ cm^{-1}), acetone ($\Delta \nu = 3020$ cm^{-1}), a mixture of acetone and carbon disulfide, and in calcite crystals ($\Delta \nu = 1085$ cm^{-1}).

The SRS was excited by a laser operating in single-mode conditions in the experimental setup shown in Fig. 1 (see Section 5). The investigations of the angular distribution of the first Stokes component of benzene were carried out in cells 10 and 40 cm long. The energy of the exciting radiation ranged from 0.032 to 0.027 J. Figure 5 shows the spectrum of SRS in benzene at a cell length 40 cm, while Fig. 6 shows the results of the photometry of the angular distribution in benzene at a cell length 10 cm.

The results of the measurements of the half-widths of the angular distribution of the exciting radiation passing through the medium and of the first Stokes components are given in Table 2 for a cell 40 cm long.

The average measurement error was $\pm 1 \cdot 10^{-3}$ rad. The scatter in the half-widths of the angular distribution is determined by the photometry errors and by the instability of the SRS excitation conditions. To estimate the photometry errors we have performed control measurements of the angular distribution with a lens of focal length 1000 mm. Since the increase of the focal length of this lens did not lead to a noticeable decrease of the measurement errors, the rest of the study was performed under the initial conditions.

Measurements of the half-width Δ_{SRS} of the angular distribution of the first SRS Stokes component in benzene, using a single-mode laser and a cell 40 cm long, yielded $\delta_{SRS} = 0.004 \pm 0.001$ rad.

For the half-width Δ_R of the angular distribution of the exciting radiation passing through the cell we obtained correspondingly the value $\delta_R = 0.003 \pm 0.001$ rad.

When SRS was excited in a cell 10 cm long, the angular distribution of the first Stokes component turned out to agree, within the limits of errors, with the distribution in a cell 40 cm long, namely $\delta_R = 0.003$ rad and $\delta_{SRS} = 0.0045$ rad.

Thus, changing the cell length does not affect the angular distribution of the first Stokes component.

As can be seen from the presented data, the angular distribution of the first Stokes component of the SRS, using a single-mode laser, differs insignificantly from the angular distribution of the exciting radiation, i.e., it approximately "duplicates" this distribution.

If a multimode laser is used, the picture of the angular distribution of the first Stokes component of the SRS becomes more complicated. Namely, the scattering spot is split in many cases into several small dots usually located along the spectral line. The total width of this complicated distribution is sometimes 10–15

Fig. 5. Spectrum of SRS in benzene under single-mode excitation. Exc — exciting radiation passing through the medium; 1St — first Stokes component.

times larger than the width of the ruby-radiation distribution. In the direction perpendicular to the arrangement of the small spots (along the diameters of the separate spots) the widths of the angular distribution of the SRS and of the ruby radiation differ less. By way of illustration, Fig. 7 shows the obtained distributions for acetone. A quantitative comparison of the angular distribution of the ruby radiation passing through the cell and the first SRS Stokes component, as indicated above, was effected by photometry of the photographs. The photometry of the compared photographs of the ruby radiation and of the SRS was carried out in the same direction on the photographic plate, for example, perpendicular to the line joining the small spots of a structure.

To be able to trace the influence of the self-focusing on the angular distribution of SRS excited by a multimode laser, we investigated several substances in which self-focusing took place (carbon disulfide, benzene), and substances in which no self-focusing took place under our excitation conditions (acetone, calcite crystals). In addition we investigated a mixture of 92% acetone and 8% carbon disulfide. According to the published data, self-focusing takes place readily in this mixture, a fact very distinctly manifest in the structure of the SRS emission in the "near field" (see [77]). By comparing the angular distribution of the first SRS Stokes component in pure acetone and in a mixture of acetone with CS_2, it was easy to trace the influence of self-focusing on the investigated angular distribution.

The results of the measurements of the half-width of the angular distribution in various media are shown in Table 3. The investigations were made in a cell 30 cm long.

It can be seen that the angular distribution of the first Stokes component of SRS excited by a laser with a complex mode composition becomes broader than that of the exciting radiation passing through the cell. However, the width of the angular distribution of the first Stokes component still turns out to be much less than expected from the geometrical conditions of the experiment (see Section 10 below). In the case of scattering in calcite it is necessary to take into account the broadening of the angular distribution as a result of the inhomogeneity of the medium. As follows from Table 3, self-focusing does not influence substantially the width of the angular distribution of the first Stokes component.

The data obtained by us allows us to deduce the presence of a "replica effect" in the angular distribution of the first Stokes component of SRS. Namely, the angular distribution of this component coincides, in first-order approximation, with the angular distribution of the exciting radiation passing through the scattering medium.

According to the data given above, the replica effect manifests itself most strongly when a single-mode laser is used. The use of a multimode laser, the presence of inhomogeneities in the medium, as well as other interfering factors can cause the replica affect to become masked by extraneous phenomena.

8. Angular Distribution of Paraxial Radiation of Higher Stokes Components

The investigations of the angular distribution of paraxial radiation of higher Stokes components were carried out for the SRS spectra of several organic liquids (benzene, carbon disulfide, nitrobenzene).

We used for the measurements the same setup and procedures as in the study of the angular distribution of the first SRS Stokes component (see Section 5). In view of the difficulty of exciting the higher Stokes SRS components at low exciting-radiation power, we were unable to perform these measurements with a single-mode laser. We used instead a laser operating with one axial mode and several angular modes. As a result, the width of the angular distribution of the exciting radiation entering this cell with the investigated substance was approximately 0.01 rad.

The measured values of the half-widths of the angular distribution are listed in Table 4. The column marked "Ruby" gives the data for the angular distribution of the exciting substance.

TABLE 2. Half-Widths of the Angular Distribution of the Exciting Radiation Passing through the Medium and of the First SRS Stokes Component in Benzene for Single-Mode Laser Excitation

Photograph	δφ, 10⁻³ rad		Photograph	δφ, 10⁻³ rad	
	exciting radiation passing through the medium	first Stokes component		exciting radiation passing through the medium	first Stokes component
1	4.6	6.8	3	1.7	2.6
2	3.6	5.6	4	2	3.3

TABLE 3. Half-Width of the Angular Distribution in Direction Perpendicular to the Spectrum

Substance	δφ, 10⁻¹ rad		Substance	δφ, 10⁻² rad	
	exciting radiation	first Stokes component		exciting radiation	first Stokes component
Acetone	2.4	5.6	Benzene	3.2	6.7
Mixture of acetone with CS_2	3.2	6.7	Calcite	3.5	8.7

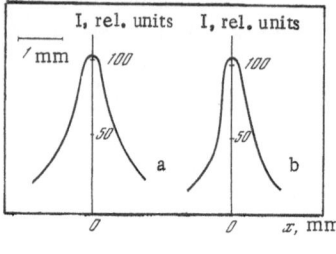

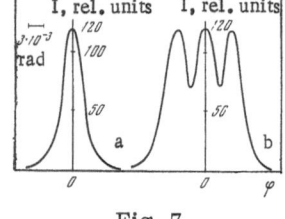

Fig. 6 Fig. 7

Fig. 6. Distribution of the intensity of the radiation for benzene excited by a single-mode laser. (a) First Stokes component of SRS; (b) ruby radiation passing through the cell (the dimensions of the spots on the photographic plate are indicated).

Fig. 7. Angular distribution of the intensity of the first SRS Stokes component in acetone following excitation by a multi-mode laser. (a) Perpendicular arrangement of the small spots of the structure; (b) along the arrangement of the small spots.

Figure 8 shows the measurement results graphically. As can be seen, the half-width of the angular distribution of all the SRS Stokes components in all the investigated substances is the same within the limits of the experimental error (approximately 15%), and coincides in practice with the half-width of the angular distribution of the exciting radiation passing through the cell. Thus, the "replica effect" takes place for the angular distribution of all the Stokes components investigated by us.

The results can be easily understood from the point of view of the stepwise mechanism of SRS excitation, according to which the exciting line of the second Stokes component propagating along the axis of the light beam is the first Stokes component. Thus, the first SRS Stokes component "duplicates" the angular distribution of the exciting radiation, and the second Stokes component "duplicates" the angular distribution of the first Stokes component.

This process extends also to higher components, which are obtained one after the other in the successive stages of the SRS excitation.

9. Theory of Stokes — Stokes Coherent Four-Photon Processes in SRS

In the present paper we interpret the experimental results from the point of view of the theory developed by Sushchinskii [7, 107, 108] for coherent processes in SRS. These processes manifest themselves particularly

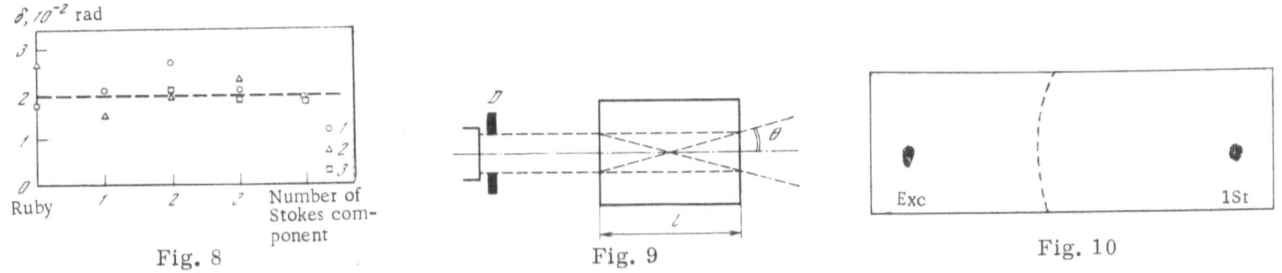

Fig. 8

Fig. 9

Fig. 10

Fig. 8. Half-widths of the angular distribution for various Stokes components. (1) Carbon disulfide; (2) nitrobenzene; (3) benzene.

Fig. 9. Distribution of the radiation of the first Stokes SRS component with respect to the angles in the case of two-photon processes. Cell length $l = 100$ mm, beam cross-sectional diameter $D = 5$ mm, $\theta = 0.05$ rad.

Fig. 10. SRS spectrum in benzene. Exc − exciting radiation passing through the cell with the substance; 1St − first Stokes component.

TABLE 4. Half-Width of the Angular Distribution of the Paraxial Radiation of Various SRS Components (in Units of 10^{-2} rad)

Substance	Ruby	Stokes components of SRS			
		1St	2St	3St	4St
Carbon disulfide	1.7	2,0	2,3	2,0	1,8
Nitrobenzene	2,6	1,5	1,9	2,3	—
Benzene	—	—	2,0	1,7	1,7

strongly in four-photon light scattering, which leads to the appearance of radiation rings of anti-Stokes and higher Stokes SRS components. Four-photon scattering in SRS is subject to definite phase relations, and its intensity is of the same order of magnitude as ordinary two-photon scattering. It is therefore natural to regard this scattering not as a fourth-order process, but as a coherent second-order process (see [108]).

The question of the physical causes of the indicated coherence is not considered in our paper. This problem is the subject of papers by Makhviladze and Shelepin [116–119].

We note that in ordinary analysis of Raman-scattering processes [7, 120], the phases of the incident and scattered photons are of no importance and can be disregarded. Such an approach will not do in the investigation of four-photon processes, in which two photons are incident on the molecule and two new photons are scattered. In this case, besides the ordinary (noncoherent) scattering processes, an important role is played by coherent processes that are subject to definite phase relationships.

Following [108], we consider two processes A and B with respective amplitudes ψ_a and ψ_b. If these processes are independent, then the total probability of the transition C = A + B is equal to the sum of the probabilities of the two indicated transitions − the processes are noncoherent:

$$W_{noncoh} = \psi_a \psi_a^* + \psi_b \psi_b^* = W_A + W_B. \tag{9.1}$$

In the analysis of SRS in [108], the author introduces the assumption that the amplitudes of the second-order processes are summed. Accordingly, the amplitude ψ of a complex coherent process is equal to

$$\psi = \psi_a + \psi_b, \tag{9.2}$$

and for the probability of the coherent process we have

$$W_{coh} = \psi_a \psi_a^* + \psi_b \psi_b^* + \psi_a \psi_b^* + \psi_b \psi_a^* = W_{noncoh} + \Delta W. \tag{9.3}$$

The process under consideration is actually coherent only if $\Delta W \neq 0$.

We apply these arguments to the two Raman light-scattering processes, and use the notation and results of [7]. We assume that in processes A the molecule goes over from state i to state k, with absorption of the photon of frequency ω_a and wave vector $\mathbf{k}_a$, and emission of a photon with frequency ω_a' and wave vector $\mathbf{k}_a^i$.

21

In process B the initial and final states of the molecule are the same as in process A, while the frequencies and wave vectors of the absorbed and emitted photons are, respectively, ω_b, $\mathbf{k}_b$ and ω_b', $\mathbf{k}_b'$.

The amplitude of process A is equal to

$$\psi_a = \frac{2\pi f_{ki}(t)\sqrt{\omega_a \omega_a'}\sqrt{n_a\left(n_a' + \frac{\omega_a'^2}{(2\pi c)^3}\right)}}{|\,\omega_a' + \omega_I - \omega_a + i\,(q_i - q_k)\,|}\,|\,S_{ki}^{(a)}\,|\,e^{i(\mathbf{k}_a - \mathbf{k}_a')\mathbf{r}}\,e^{i(\varphi_a + \varphi_a')}, \tag{9.4}$$

where

$$f_{ki}(t) = e^{-q_k t} - e^{-q_i t}\,e^{i(\omega_a' + \omega_I - \omega_a)t}, \tag{9.5}$$

$$S_{ki}^{(a)} = \sum \frac{(e_a M_{li})(e_a' M_{kl})}{\omega_l^e - \omega_a + i\,(q_i - q_l)} + \frac{(e_a' M_{li})(e_a M_{kl})}{\omega_l^e + \omega_a' + i\,(q_i - q_l)}. \tag{9.6}$$

In these equations q_i and q_k are the widths of the corresponding levels; ω_I is the frequency of the vibrational transition; e_a and e_a' are the polarization vectors of the absorbed and emitted photons; M_{li} and M_{kl} are the matrix elements of the dipole moments; $\hbar\omega_l^e$ are the energies of the intermediate electronic states; n_a and n_a' are the numbers of the incident and scattered photons; the constant phase factors $e^{i\varphi_a}$, $e^{i\varphi_a'}$ take into account, respectively, the imaginary terms of the denominators of (9.6) and (9.4).

The expressions for the amplitude of the process B are similar.

We denote by $\rho(\omega_0 - \omega)$ and $\rho(\Omega_0 - \Omega)$ the distributions of the amplitudes of the incident radiation with respect to the frequencies and the angles (by Ω is meant the entire aggregate of the angle variables). With the aid of (9.4) and the analogous expression for the process B we obtain $\psi_a \psi_b^* + \psi_b \psi_a^*$. This expression will contain the oscillating factor $\cos(\Delta\omega t + \Delta\mathbf{k}\cdot\mathbf{r} + \Delta\varphi + \Delta\varphi')$. In the case of Stokes–Stokes processes $q_i = 0$, $q_k = q$ and at large qt the first term in (9.5) can be neglected.

To calculate ΔW we must integrate the obtained expressions with respect to the frequencies and angles of the exciting lines.

After simple calculations we obtain

$$\Delta W = a\sqrt{n_a n_b}\,F(\omega_a', \omega_b', t)\,\Phi(\Omega_a', \Omega_b', r)\cos(\Delta\varphi + \Delta\varphi'), \tag{9.7}$$

$$F(\omega_a', \omega_b', t) = \int \frac{\rho(\omega_{0a} - \omega_a)\,\rho(\omega_{0b} - \omega_b)}{|\,\Delta\omega_a - iq\,|\,|\,\Delta\omega_b - iq\,|}\cos(\Delta\omega t)\,d\omega_a\,d\omega_b, \tag{9.8}$$

$$\Phi(\Omega_a', \Omega_b', r) = \int \gamma(\mathbf{k}, \mathbf{r})\cos(\Delta\mathbf{k}\cdot\mathbf{r})\,d\Omega_a\,d\Omega_b,$$

$$a = 8\pi^2\sqrt{\omega_a \omega_a' \omega_b \omega_b'}\sqrt{n_a' + \frac{\omega_a'^2}{(2\pi c)^3}}\sqrt{n_b' + \frac{\omega_b'^2}{(2\pi c)^3}},$$

$$\gamma(\mathbf{k}, \mathbf{r}) = |\,S_{ki}^{(a)}\,|\,|\,S_{ki}^{(b)}\,|\,\rho(\Omega_{0a} - \Omega_a)\rho(\Omega_{0b} - \Omega_b), \tag{9.9}$$

$$\Delta\omega_a = \omega_a' + \omega_I - \omega_a; \quad \Delta\omega_b = \omega_b' + \omega_I - \omega_b; \quad \Delta\omega = \Delta\omega_b - \Delta\omega_a.$$

At $\Delta\omega \neq 0$ and $\Delta\mathbf{k} \neq 0$ the integrals (9.8) and (9.9) constitute rapidly decreasing functions of the time and of the spatial coordinates, respectively. If, for example, $\rho^2(\omega_{0a} - \omega_a)$ takes the form of a dispersion or Gaussian function, then we have correspondingly $F \sim e^{-\delta t}$. Thus, at large T, generally speaking, $\Delta W = 0$ and the considered processes are noncoherent.

The quantity ΔW can differ from zero only if the region of integration in (9.8) and (9.9) is defined by the conditions

$$\Delta\omega = \omega_a - \omega_a' - \omega_b + \omega_b' = 0, \tag{9.10}$$

$$\Delta\mathbf{k} = \mathbf{k}_a' - \mathbf{k}_a - \mathbf{k}_b + \mathbf{k}_b' = 0. \tag{9.11}$$

The wave vectors expressed in this form represent the coherence relation of the considered Stokes–Stokes processes. We note that (9.11) is, in somewhat modified form, the well-known phase-synchronism condition.

Expression (9.7) shows that the process A is induced not only by the photons that participate in this process, but also by the photons that participate in process B, and vice versa. We can regard this expression as defining the probability of the "four-photon" Raman scattering process, since the wave vectors of the photons are connected by relations (9.10) and (9.11). It follows from the foregoing analysis that the probability of this process differs from zero only when the coherence conditions are satisfied.

The considered coherent Raman scattering is added to the ordinary noncoherent Raman scattering. In the case of SRS this leads to a predominant development of processes that satisfy the coherence laws.

10. Discussion of the Results on the Angular Distribution

of Paraxial SRS Radiation

The systematic study of the angular distribution of the paraxial SRS radiation, reported in the present paper, shows that this distribution is characterized by a strongly pronounced forward directivity. This directivity manifests itself in the fact that the angular distribution of the first Stokes component of the SRS "duplicates" so to speak the distribution of the exciting radiation. Of course, the angular distribution of the SRS "duplicates" the corresponding distribution of the exciting radiation only with some approximation. This is quite natural if it is recognized that processes that lead to a broadening of the indicated distributions are always present.

To explain the predominant directivity of the first Stokes component of the SRS along the axis of the exciting radiation, arguments are usually advanced concerning the differences between the path lengths l along the laser beam and away from it. Since the intensity of the SRS lines depend exponentially on l, this factor must of course be taken into account. However, a simple geometrical consideration leads to the conclusion that it is impossible to explain the experimental data in this manner.

The simplest to analyze is the case when a parallel beam of exciting radiation passes through the cell with the investigated substance (Fig. 9). The beam entering the cell contains only photons having the frequency ω of the laser. These photons excite at first ordinary Raman scattering. This process is the only real source of the primary photons at the Stokes frequency ω'. The angular distribution of these photons in the angle interval of interest to us (near the axis) can be regarded as almost uniform.

We follow next the usual amplification scheme. Each Stokes photon induces photons in the same direction as that of the photon under consideration. But under the identical conditions that are obtained near the axis, the amplification should be the same in all directions. The cross section of the beam of the exciting radiation along the cell axis is in our case rectangular in shape. The length of the light path at small angles θ to the axis is obviously larger than along the cell axis. Accordingly, the intensity of the SRS at small angles θ to the axis, for example along the diagonal of the beam cross section, should not be less than along the axis.

The experimental data strongly contradict this conclusion. By way of example, the dashed line in Fig. 10 shows a part of the circle in which the uniform SRS radiation should be distributed in accordance with the presented calculation, as well as the experimentally obtained distribution.

It should be noted that the discrepancy between the experiment and the ordinary SRS theory, observed in our studies, is not confined to the angular distribution of the SRS radiation. A similar discrepancy occurs in the spectral width of the SRS radiation, which turns out, for the first Stokes component, to be much narrower than expected from the theory of the phenomenon [121, 122]. In accordance with the hypothesis advanced in [108], both "replica effects" are closely connected and can be explained on a common basis.

The detailed analysis presented above (Section 9) shows that the SRS intensity in coherent processes differs from zero only when the following coherence conditions are satisfied:

$$\Delta \mathbf{k} = \mathbf{k}_a - \mathbf{k}'_a - \mathbf{k}_b + \mathbf{k}'_b = 0, \tag{10.1}$$

$$\Delta \omega = \omega_a - \omega'_a - \omega_b + \omega'_b = 0. \tag{10.2}$$

We consider the particular case when both incident photons are photons of the exciting radiation. The coherent process of this type is superimposed on the ordinary SRS process that leads to formation of the first Stokes component. For this coherent process the condition $\Delta \omega = 0$ leads to the conclusion that at $\omega_a - \omega_b = \delta \omega$ we also have $\omega'_a - \omega'_b = \delta \omega$. But this means that the width of the SRS line in coherent scattering coincides with the width of the exciting line, i.e., we have here the "repetition" effect with respect to the line width. From the condition (10.2) we find similarly that if $\mathbf{k}_a - \mathbf{k}_b = \delta \mathbf{k}$, then

$$\mathbf{k}'_a - \mathbf{k}'_b = \delta \mathbf{k}. \tag{10.3}$$

This means that the angular divergence of the scattered radiation does not extend beyond the limits of the divergence of the exciting radition, i.e., the "replica effect" with respect to angles, as observed in our experiments, does take place. Coherent (four-photon) Raman scattering is superimposed on the ordinary noncoherent (two-photon) Raman scattering (the probabilities of these two processes are approximately equal). In the case of SRS this leads to a predominant development of processes that satisfy the coherence conditions.

If the coherence conditions cannot be satisfied (for example, in an off-axis resonator), then processes of amplification of the noncoherent SRS develop. The transition to the saturation region leads to an increase in the field of the noncoherent SRS, a fact that should manifest itself in a broadening of the spectral line and of the angular distribution of the SRS.

In purest form, the repetition effects should be observed at small excess above the SRS threshold.

CHAPTER IV

INVESTIGATIONS OF OFF-AXIS CLASS-II SRS EMISSION

11. Results of Measurements of the Angular Distribution
of Class-II Off-Axis SRS Emission

We have investigated class-II emission in three organic liquids: benzene, carbon disulfide, and nitrobenzene.

We used for the investigations the setup described in detail in Section 5 (a diagram of the setup is shown in Fig. 3). In this setup, the SRS was excited first in the auxiliary cell 9 filled with the investigated liquid. In accord with the experimental conditions, the first and second SRS Stokes components were excited in the auxiliary cell.

The second SRS Stokes component, when necessary, was attenuated with selective filters. The spectral composition of the radiation emerging from the auxiliary cell was monitored photographically in spectrograph 18 (see Fig. 3).

The off-axis SRS emission was excited in the main cell 15, which ranged in length from 10 to 30 cm. The radiation propagating along the axis, after emerging from the auxiliary cell 9, was directed into the main cell 15 by lens 14. The off-axis radiation emerging from cell 9 was cut off by diaphragm 11. Thus, the conditions of the excitation of the SRS in cell 15 were chosen such that only off-axis class-II radiation could be produced in it. A characteristic feature of the radiation of this class is that the angle between this radiation and the axis is independent of the focal length of the focusing lens 14. In our experiments, when the focal length of this lens was changed from 12 to 110 cm the diameters of the observed rings remain unchanged. The radiation emerging from cell 15 was focused by lens 16 on the plane of the slit of the spectrograph 17 (the slit was removed in the course of the experiment). The plane of the slit coincided with the focal plane of the lens 16, i.e., we investigated the distribution of the SRS intensity in the "far field."

In the excitation of the second SRS Stokes component that propagates along the axis, the first Stokes component plays the role of the exciting radiation. Analogously, in the excitation of the axial radiation of the third Stokes component, the role of the exciting radiation is played by the second Stokes component propagating along the axis, etc. The conditions for the excitation of the off-axis SRS radiation, in which an important role is played by four-photon processes, are more complicated. As will be shown below, when class-II off-axis radiation is excited, the initial process is the one in which the ruby radiation and the radiation of the first Stokes SRS component, propagating along the axis, are transformed into off-axis radiation of the second Stokes and the first anti-Stokes SRS components.

Both axial and off-axis radiation of various SRS components can participate in the subsequent four-photon processes.

In the setup employed by us we were able to direct into the main cell the axial radiation of the ruby and of the first Stokes SRS component of desired intensity. The practical conditions of the measurements were chosen such that at constant ruby-radiation intensity it was possible to vary, in a wide range, the intensity of the first Stokes component emerging from the auxiliary cell. We have thus investigated the intensity of the second Stokes component of the SRS and the distribution of the intensity between the "central spot" and the "ring" as a function of the intensity of the first Stokes SRS component.

We have investigated also the axial and off-axis radiation of the third and fourth Stokes component and of the first anti-Stokes component of the SRS. As indicated above (see Section 6), in our setup it was easy to obtain sufficiently intense higher SRS Stokes components. In those cases when the dependence of the intensity

TABLE 5. Results of Measurement of the Angles of Class-II Off-Axis SRS Radiation

Substance	SRS components	Angles, rad		
		from the data of [25]	from the data of [123]	our data
Benzene	1St	0.0325	0.0313	0.035±0.004
	2St	0.038	—	0.038±0.005
	3St	—	—	0.037±0.005
	4St (θ_{-4})	—	—	0.036±0.006
	4St (θ'_{-4})	—	—	0.051±0.007
Carbon disulfide	2St	—	—	0.030±0.004
	3St	—	—	0.034±0.004
	4St	—	—	0.033±0.005
Nitrobenzene	2St	0.065	—	0.073±0.007
	3St	—	—	0.072±0.007

of the off-axis radiation of the third SRS Stokes component on the intensity of the second Stokes component propagating along the axis was investigated, the latter was not attenuated prior to its entry into the working cell.

The results of the measurements of the angular distribution of class-II radiation are listed in Table 5. It is seen that the data obtained by us are in satisfactory agreement with the published data on the first anti-Stokes and the second Stokes components (all listed angles are in air).

Measurements were made of the off-axis radiation angles for higher Stokes components, for which there are practically no published data.

The off-axis class-II radiation produced, on the photographs, single rings for the second and third Stokes components and for the first anti-Stokes component. The fourth Stokes component consisted of two concentric rings with angles θ_{-4} for the inner ring and θ'_{-4} for the outer.

The diameters of the rings varied somewhat from photograph to photograph; this led to an uncontrollable error in the measurement of the angles. Thus, despite all the precautions taken in the measurements of the angles and the large statistics, the error in the angle values given below is 10-15%.

The ratio of the intensities of various types of SRS radiation for the higher Stokes components is discussed in Section 14.

12. Structure of Class-II SRS Radiation Rings

It is frequently indicated in [25] that the class-II SRS rings have a large width and are diffuse. On the contrary, our investigations of the Stokes components revealed more frequently rather sharp and narrow class-II rings. We have therefore investigated in detail the structure of the class-II rings as a function of the SRS excitation conditions. We have similarly investigated the cross-sectional intensity distribution of the SRS Stokes components — the profile of the class-II rings.

We investigated the profile of the rings of the second, third, and fourth SRS Stokes components in benzene. The intensities were measured by the method of photographic photometry described in Section 5. In the photometry, the distance between the individual points on the photograph ranged from 0.1 to 0.25 mm, which made it possible to trace sufficiently reliably the variation of the intensity in the cross sections of the rings.

Figure 11 shows typical profiles of the second Stokes component at various ratios of the intensities of the paraxial and off-axis radiation in benzene. The measured values of the half-widths of the angular distribution of the off-axis radiation δ_r and of the paraxial radiation δ_c are listed in Table 6.

As shown by the results, the half-width δ_r of the class-II rings of the second and third SRS components is small. It is equal to approximately one-half the half-width δ_c of the axial SRS emission, i.e.,

$$\delta_r = \frac{1}{2}\delta_c.$$

A noticeable broadening of the profile of the rings is observed on going to the fourth Stokes component. This broadening is due to the more complicated structure of the off-axis radiation of the fourth Stokes SRS component. A number of photographs obtained with the fourth component in benzene revealed two concentric

TABLE 6. Half-Width of the Profile of the Angular Distribution of Various SRS Components in Benzene (in Units of 10^{-2} rad)

Characteristic of radiation	Ruby	2St	3St	4St
Axial radiation	2,1	2.0	1.7	1.7
Off-axis radiation	—	1.0	1.1	2.2

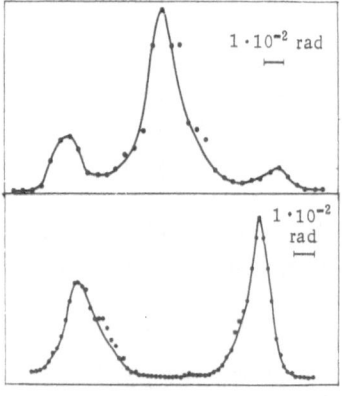

Fig. 11. Radiation profiles of class-II rings of the second SRS Stokes component in benzene at various intensity ratios of the paraxial and off-axis radiation.

rings. In those cases when the rings overlapped partially, they produced a combined ring with a broader profile.

Besides the described symmetrical structure of the rings, many photographs reveal asymmetrical ring distortions. In the simplest case they manifest themselves in unequal intensities and widths of the rings. Such asymmetrical distortions of the rings can be naturally attributed to the inhomogeneity of the angular distribution of the radiation in the exciting beam.

13. Theory of Angular Distribution of Off-Axis Class-II

SRS Emission

The data obtained on the angular distribution of the SRS class-II emission can be explained on the basis of the theory of coherent four-photon processes, developed in [105-110]. This theory elucidates the conditions under which two Raman-scattering processes A and B with respective amplitudes ψ_a and ψ_b are coherent, i.e., their probability

$$W = \psi_a\psi_a^* + \psi_b\psi_b^* + \psi_a\psi_b^* + \psi_b\psi_a^* \tag{13.1}$$

has a nonzero component $\Delta W = \psi_a\psi_b^* + \psi_b\psi_a^*$. The general theory proposed in Chapter III as applied to Stokes − Stokes processes can be easily extended to Stokes − anti-Stokes processes. These processes, as will be shown below, explain completely all the characteristics of the radiation of class-II without resorting to any additional assumptions. Figure 12 shows the scheme of transitions between the unexcited vibrational level v = 0 and the excited vibrational level v = 1. Each of them consists of two simple two-photon Raman-scattering processes.

For the amplitudes of the indicated processes we can use the known expressions (see [7])

$$\psi_a = a_0 e^{i(\mathbf{k}_a - \mathbf{k}_a')\mathbf{r}}, \qquad \psi_b = b_0 e^{i(\mathbf{k}_b - \mathbf{k}_b')\mathbf{r}}, \tag{13.2}$$

where $\mathbf{k}_a$ and $\mathbf{k}_b$ are the wave vectors of the incident photons, and $\mathbf{k}_a'$ and $\mathbf{k}_b'$ are those of the scattered photons in processes A and B, respectively. In order for ΔW not to vanish upon averaging over the different molecules, the processes A and B must take place simultaneously on one and the same molecule. If A and B are both Stokes processes, then the amplitude of the four-photon process is $\psi_{St\ St} = \psi_a + \psi_b$. In this case

$$\Delta W = 2a_0b_0 \cos[(\mathbf{k}_a - \mathbf{k}_a' - \mathbf{k}_b + \mathbf{k}_b')\mathbf{r}]. \tag{13.3}$$

In the integration with respect to $\mathbf{r}$, the quantity ΔW does not vanish only under the condition (see Section 9)

$$\mathbf{k}_a - \mathbf{k}_a' - \mathbf{k}_b + \mathbf{k}_b' = 0. \tag{13.4}$$

26

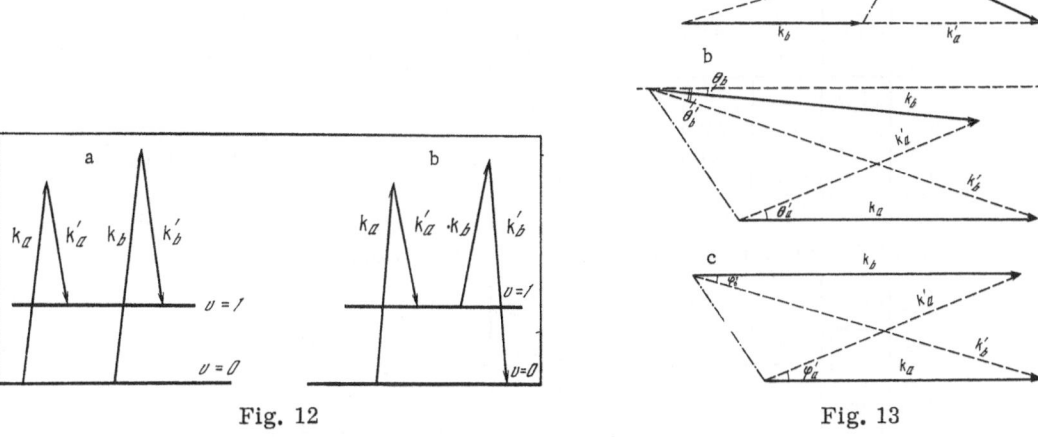

Fig. 12   Fig. 13

Fig. 12. Transition schemes. (a) Stokes—Stokes transitions; (b) Stokes—anti-Stokes transition.

Fig. 13. Wave-vector arrangement corresponding to different phase-synchronism conditions.

This relation is the usual phase-synchronism condition. It is satisfied for the wave-vector arrangement shown in Fig. 13a (the vectors of the incident photons are shown by solid lines and those of the scattered photons are shown dashed).

If one of the two processes is a Stokes process and the other anti-Stokes, then the coherence conditions are somewhat modified. In this case, since the matrix elements for the Stokes and anti-Stokes transitions are Hermitian adjoints (see [7]), the amplitude of the process B is

$$\psi_b = b_0 e^{-i[(k_b - k_b')r]}$$

and consequently $\psi_{St\,a} = \psi_a + \psi_b^*$. We obtain correspondingly the coherence condition for the wave vectors in Stokes—anti-Stokes transitions in the form ($k_a \neq k_b$)

$$\mathbf{k}_a - \mathbf{k}_a' - \mathbf{k}_b' + \mathbf{k}_b = 0. \tag{13.5}$$

This condition corresponds to the wave-vector arrangements shown in Fig. 13b. For the angle θ_b' between the vector $\mathbf{k}_b'$ and the direction of the vector $\mathbf{k}_a$ we obtain after straightforward rather cumbersome calculations

$$\sin(\theta_b' - \theta_b) = -\frac{k_a \sin\theta_b (Q^2 + k_b'^2 - k_a'^2)}{2k_b' Q^2} +$$

$$+ \frac{(k_b + k_a \cos\theta_b)}{2k_b' Q^2}[(k_b' + k_a' - Q)(Q + k_a' - k_b')(Q + k_b' - k_a')(Q + k_a' + k_b')]^{1/2}. \tag{13.6}$$

Here $Q^2 = k_a^2 + 2k_a k_b \cos\theta_b + k_b^2$; θ_b is the angle between the direction of the vectors $\mathbf{k}_b$ and $\mathbf{k}_a$. In the important particular case when the vectors $\mathbf{k}_a$ and $\mathbf{k}_b$ are parallel (Fig. 13c), Eq. (13.6) becomes much simpler. Designating for this case the angles that the vectors $\mathbf{k}_a'$ and $\mathbf{k}_b'$ make with the axis by φ_a', φ_b', respectively, we have

$$\sin\varphi_b' = \frac{[\Pi(k_a + k_b + k_b' - k_a')(k_a + k_b + k_a' + k_b')]^{1/2}}{2k_b'(k_a + k_b)}, \tag{13.7}$$

where $\Pi = (k_b' + k_a' - k_b - k_a)(k_a + k_b + k_a^* - k_b')$,

$$\sin\varphi_a' = (k_b'/k_a')\sin\varphi_b'. \tag{13.8}$$

We note that in Stokes—anti-Stokes transitions the anti-Stokes radiation is produced by four-photon scattering from unexcited molecules. The intensity of this radiation can therefore be quite appreciable.

The evolution of the SRS process with participation of Stokes—anti-Stokes transitions can be visualized as follows. Initially only SRS at the first Stokes frequency, with wave vector $\mathbf{k}_a = \mathbf{k}_{-1}$ parallel to the vector of the exciting radiation $\mathbf{k}_b = \mathbf{k}_0$, is present in the investigated substance (cell 15 of Fig. 3). These two vectors are the starting points in the Stokes—anti-Stokes transition (Fig. 13c) with formation of the second Stokes component

TABLE 7. Off-Axis SRS Emission Angles in Benzene, rad

SRS components	Calculation	Experiment			SRS components	Calculation	Experiment		
		[25]	[123]	our data			[25]	[123]	our data
3aSt	0,0810	0,085	0,0853	—	3St	0,0435	—	—	0,037±0,005
2aSt	0,0568	0,060	0,0553	—	4St	$\begin{cases} 0,0436 \\ 0,0553 \end{cases}$	--	—	0,036±0,006 / 0,051±0,007
1aSt	0,0348	0,0325	0,0313	0,035±0,004					
2St	0,0433	0,038	—	0,038±0,005					

TABLE 8. Off-Axis SRS Angles in Carbon Disulfide, rad

SRS components	Calculation	Experiment			SRS components	Calculation	Experiment		
		[25]	[123]	our data			[25]	[123]	our data
3aSt	0,0812	0,080	0,078	—	2St	0,0412	—	—	0,030±0,004
2aSt	0,0580	0,051	0,050	--	3St	0,0409	—	—	0,034±0,004
1aSt	0,0358	0,033	—	—	4St	0,0406	—	—	0,033±0,005

TABLE 9. Off-Axis SRS Emission Angles in Nitrobenzene, rad

SRS components	Calculation	Experiment			SRS components	Calculation	Experiment		
		[25]	[123]	our data			[25]	[123]	our data
2aSt	0,100	0,101	—	—	2St	0,0760	0,065	—	0,073±0,007
1aSt	0,0566	0,058	0,0569	—	3St	0,0756	—	—	0,072±0,007

$\mathbf{k}'_a = \mathbf{k}_{-2}$ and the first anti-Stokes component $\mathbf{k}'_b = \mathbf{k}_1$. At **sufficiently high** intensity of the first anti-Stokes component propagating at an angle $\varphi'_b = \theta_b$ to the axis, this component and the exciting radiation $\mathbf{k}_0$ in the new Stokes−anti-Stokes transition produce the second anti-Stokes component $\mathbf{k}'_b = \mathbf{k}_2$ and the first Stokes component $\mathbf{k}'_a = \mathbf{k}_{-1}$, which propagate at angles θ'_a and θ'_b to the axis (Fig. 13b). In the same way, the components $\mathbf{k}_2$ and $\mathbf{k}_0$ produce the third anti-Stokes component, etc.

The starting point in the formation of the class-II higher-order Stokes components is the axial radiation produced by successive excitation of the SRS Stokes components. This radiation is more intense than the radiation produced in four-photon processes and propagating at angles to the axis. Therefore the formation of rings of the higher Stokes components takes place in all cases in accordance with the scheme shown in Fig. 13c. For example, the third Stokes component $\mathbf{k}'_a = \mathbf{k}_{-3}$, which propagates at an angle φ'_a to the axis, is the result of the first and second axial Stokes components ($\mathbf{k}_a = \mathbf{k}_{-2}$; $\mathbf{k}_b = \mathbf{k}_{-1}$), etc.

The results of the calculations by formulas (13.7) and (13.8) above (the angles θ_1, θ_2, θ_3, θ_4) and (13.6) (angles θ_2, θ_3) are compared with the experimental data in Tables 7-9. In view of the insufficient experimental data on the dispersion, the refractive indices for the higher Stokes components were obtained by extrapolation.

It should be noted that the wave numbers and refractive indices obtained by extrapolation cannot be regarded as completely reliable. We note also that the experimental values of the angles are likewise subject to an error of 10-15%, mainly because these angles vary noticeably from photograph to photograph. This circumstance is pointed out also in [25]. If account is taken of the angle-measurement errors and of some dispersion uncertainty due to the extrapolation, then the agreement between the calculated and experimental data can be regarded as satisfactory. It must be emphasized that in the calculation of the angles we used only data on the dispersion of the investigated substances, inasmuch as Eqs. (13.6)-(13.8) do not contain any indeterminate parameters. Therefore the agreement between the experimental and calculated data, given in Tables 7-9, can be regarded as significant evidence favoring the theory of coherent Stokes−anti-Stokes transitions, used by us to interpret the class-II off-axis radiation.

The self-focusing that accompanies the SRS in accordance with the experimental conditions in the investigated liquids can apparently be regarded in this case as a side effect. Favoring this assumption are also data on the excitation of the class-II rings in the absence of self-focusing.

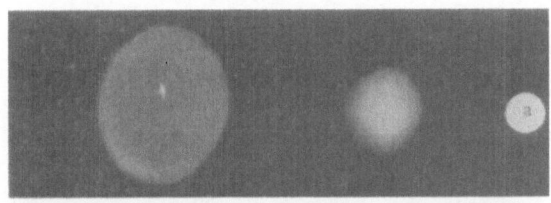

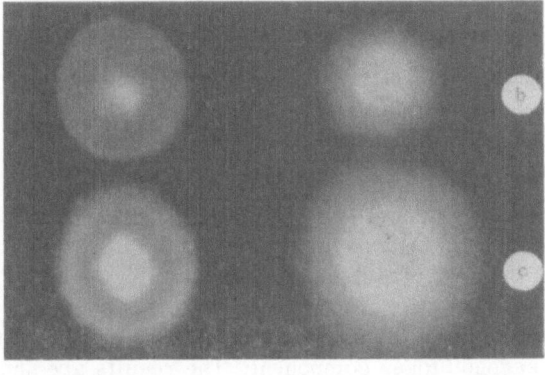

2St   1St

Fig. 14. Axial and off-axis emission of second SRS Stokes component in benzene as a function of the energy of the first Stokes component. (a) $I_{-1} = 0.02$; (b) $I_{-1} = 0.08$; (c) $I_{-1} = 0.12$ J.

14. Investigation of the Ratio of the Axial and Off-Axis

SRS Intensities

The probability of the four-photon Stokes — anti-Stokes coherent process with formation of the second Stokes component and the first anti-Stokes component, propagating at the corresponding angles to the axis, is given by the expression (see [7, 110])

$$W_{ang} = C^2 \sqrt{\omega_0 \omega_1 \omega_{-1} \omega_{-2}} \ \sqrt{n_0 n_{-1} (n_1 + g_a)(n_{-2} + g_b)}, \tag{14.1}$$

where $g_a = \omega_1^2/(2\pi c)^3$; $g_b = \omega_{-2}^2/(2\pi c)^3$; C is a constant; ω_i, n_i are the frequencies and numbers of photons of a given type of radiation (the quantities ω_0, n_0 pertain to the ruby emission, ω_{-1}, n_{-1} to the emission at the first Stokes frequency, etc.). For the emission at the second Stokes frequency, produced in the course of the subsequent excitation and propagating along the axis, we have correspondingly

$$W_{ax} = 2c^2 \omega_{-1} \omega_{-2} n_{-1} (n_{-2} + g_b). \tag{14.2}$$

It follows from the comparison of these formulas that at a low intensity of the first Stokes component, when $n_0 \gg n_{-1}$, the off-axis radiation is predominantly produced and results in class-II rings on the photographs. The condition $W_{ang} = W_{ax}$ is satisfied at

$$\frac{n_{-1}}{n_0} = \frac{1}{4} \frac{\omega_0 \omega_1}{\omega_{-1} \omega_{-2}}. \tag{14.3}$$

The intensities of the central spot and of the ring at the second Stokes frequency are then approximately equalized. Further growth of n_{-1} leads to a faster increase of the intensity of the central spot. All these conclusions are only qualitative, inasmuch as the photon numbers n_1, n_0, n_{-1}, n_{-2} change as the radiation propagates along the cell, photons of higher Stokes and anti-Stokes components are produced, etc. We note that the predominant production of off-axis radiation is possible only at $k_a \neq k_b$; at $k_a = k_b$ we always have $W_{ax} > W_{ang}$ and no rings are produced.

The conditions for the excitation of the SRS in our experiments were such that the intensity of the radiation of frequency ω_0 (i.e., with n_0 photons) entering the cell 15 (Fig. 3) remained constant, and only the intensity of the emission of frequency ω_{-1} (i.e., with n_{-1} photons) varied. The results are shown in Fig. 14. It can be seen that at low intensity of the first Stokes component I_{-1} only a ring of the second Stokes component is produced. With increasing I_{-1}, the central spot increases in intensity. The dependence of the intensity of the off-axis radiation of the second Stokes component I_{-2r} on the intensity of the first Stokes component I_{-1} is shown in Fig. 15. It agrees qualitatively with Eq. (14.1). At large values of I_{-1}, the growth of I_{-2r} slows down, apparently because of the energy transfer into other SRS components.

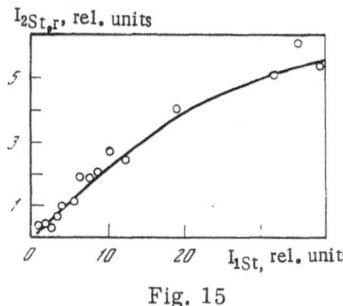
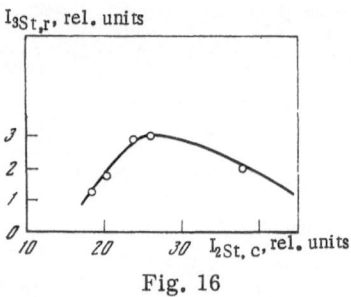

Fig. 15   Fig. 16

Fig. 15. Dependence of the intensity of the off-axis radiation of the second Stokes component of SRS of benzene on the intensity of the first Stokes component.

Fig. 16. Dependence of the intensity of the off-axis radiation of the third Stokes component of SRS in benzene on the intensity of the axial radiation of the second Stokes component.

We have investigated also the dependence of the intensity of the off-axis radiation of the third Stokes SRS component on the intensity of the axial radiation of the second Stokes component. The results are shown in Fig. 16. The forms of the curves in Figs. 15 and 16 agree at low intensities. At large values of $I_{2St.c}$, a noticeable decrease of $I_{3St.r}$ takes place. The complicated character of the obtained curve shows that at high intensities of the exciting radiation and of the Stokes components that participate in the four-photon SRS processes it is necessary to take a fuller account of all the energy transfers from one component to the other. This calls for derivation and solution of a complete system of differential equations that take into account all the transitions in the SRS, a task beyond the scope of the present paper.

As indicated in Section 11, the fourth Stokes-component emission consisted of two concentric rings. With changing intensity of the second Stokes component propagating along the axis, a change took place in the relative intensity of the rings in the radiation of the fourth Stokes component. At low intensities of the second Stokes component, only one outer ring is observed in the fourth Stokes component, with angle θ'_{-4}, and with increasing intensity an internal ring is observed with radiation at an angle θ_{-4}. The observed changes in the radiation of the fourth Stokes component of benzene are shown in Fig. 17, where the lower photographs correspond to higher intensities.

The appearance of two cones of off-axis radiation of the fourth Stokes component can be explained in the following manner. In accordance with the SRS excitation scheme described in the preceding section, one of the cones is the result of a Stokes−anti-Stokes four-photon process, in which photons with wave vectors k_{-3} and and k_{-2} (propagating along the axis) are transformed into photons k'_{-4} and k_{-1} (propagating at an angle to the axis). Calculation by this procedure yields the off-axis radiation angle θ_{-4}. A process is also possible wherein photons with wave vectors k_{-3} and k_{-1} (propagating along the axis) are transformed into photons with k'_{-1} and k'_0 (propagating at an angle to the axis). Calculation gives in this case the angle θ'_{-1} of the off-axis radiation of the fourth Stokes component. As can be seen from Table 7, the calculated angle agrees well enough with the measured angle for the outer ring of the fourth Stokes component. The formation of two rings in accord with an analogous scheme is possible in the case of the third Stokes SRS component. However, if the difference between the intensities of the axial radiation k_{-1} and k_0 is small, the processes that evolve predominantly seem to be those with participation of neighboring components. On the other hand, in the case of the fourth Stokes component the intensity of the initial second Stokes component may turn out to be less than the intensity of the first Stokes component. This in turn leads to a predominant formation of the "outer" cone of the off-axis radiation of the fourth component.

15. Calculation of Relative Intensity of Off-Axis Class-II
SRS Radiation of Various Stokes Components

It was shown above (experimentally and theoretically) that the intensity ratio of the axial and off-axis radiation of the second and higher Stokes SRS components changes significantly with the excitation conditions. At low intensity of the initial components, only off-axis radiation is excited in the four-photon SRS process, and at high intensity, on the contrary, axial radiation is produced for the main part.

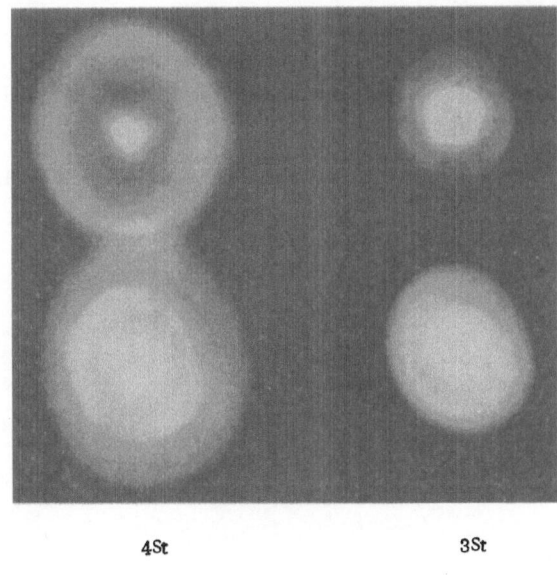

4St   3St

Fig. 17 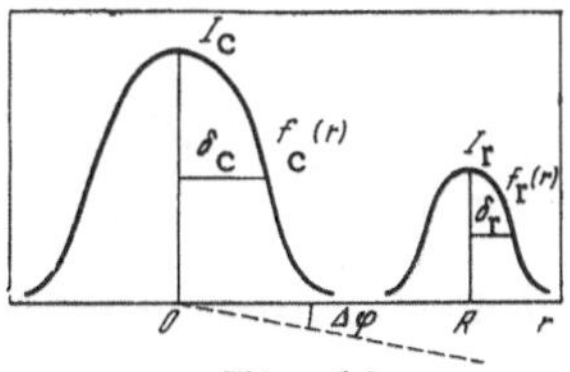   Fig. 18

Fig. 17. Axial and off-axis radiation of third and fourth components of SRS in benzene.

Fig. 18. Illustrating the calculation of the total intensity of the off-axis radiation.

Thus, the ratio of the intensity of the off-axis radiation I_r to the axial radiation I_c for each Stokes component can vary in a wide range with changing excitation conditions.

It should be noted that under certain conditions of SRS excitation, the rings of some of the higher Stokes components are not observed at all.

For example, if a sufficiently powerful axial second Stokes component enters the cell 15 (see Fig. 3) together with the first Stokes component, only the central spot of the second Stokes component was observed. This does not mean, however, that the off-axis radiation was completely absent in such cases; it manifested itself in the form of intense off-axis radiation of higher Stokes components.

In connection with the foregoing, it is of interest to determine which part of the total emission intensity of the second and higher SRS components, propagating along the axis, is made up of the off-axis radiation of these SRS Stokes components, i.e., the quantity

$$\alpha = \frac{I_r}{I_c + I_r} .$$

(15.1)

The quantity α shows which part of the radiation of the given component leaves the working channel of the system in the course of the SRS. Knowledge of this quantity is important for the dynamics of the development of the SRS, and also for the solution of practical problems connected with the operation of amplifiers and lasers based on the SRS phenomenon.

The data obtained by us allow us to conclude that when the total intensity of the Stokes component is increased, the value of α is not lower than 0.04-0.08.

For an accurate estimate of the total intensity of the off-axis radiation of the higher Stokes component it must be taken into account that this radiation is distributed on the photographs over the area of the ring, whereas the axial radiation is distributed over the area of a circle. To take into account the difference between the geometric distributions of the two types of radiation, we present the following calculation.

Assume that the radial distribution profiles of the axial radiation $f_c(r)$ and of the off-axis radiation $f_r(r)$ are described by Gaussian distribution functions:

$$f_c(r) = I_c e^{-\frac{r^2}{\delta_c^2 \ln 2}},$$

(15.2)

31

$$f_r(r) = I_r e^{-\frac{(r-R)^2}{\delta_r^2 \ln 2}}. \tag{15.3}$$

Here I_c and I_r are the intensities at the maxima of the corresponding distribution; $2\delta_c$ and $2\delta_r$ are the half-widths of the profiles (see Section 12); R is the radius of the ring.

The intensities of the axial and off-axis radiation contained in angle sectors corresponding to the angle $\Delta\varphi$ in the plane of the photograph (Fig. 18) are determined by the integrals

$$I_c' = \int_0^\infty f_c(r)\,\Delta\varphi\,r dr, \tag{15.4}$$

$$I_r' = \int_{-\infty}^\infty f_r(r)\,\Delta\varphi r dr. \tag{15.5}$$

Simple calculations yield

$$I_c' = \frac{1}{2} I_c \Delta\varphi \int_0^\infty e^{-\frac{r^2}{\delta_c^2 \ln 2}} 2r dr = \frac{1}{2} I_c \Delta\varphi \int_0^\infty e^{-\frac{z}{\delta_c^2 \ln 2}} dz,$$

where $z = r^2$. Therefore, putting $\Delta\varphi = 2\pi$, we get

$$I_c' = \pi I_c \delta_c^2 \ln 2. \tag{15.6}$$

Similarly, making the substitution $r - R = z$, we get

$$I_r' = \frac{1}{2} I_r \Delta\varphi \int_{-\infty}^\infty e^{-\frac{(r-R)^2}{\delta_r^2 \ln 2}} 2r dr = \frac{1}{2} I_r \Delta\varphi \int_{-\infty}^\infty e^{-\frac{z^2}{\delta_r^2 \ln 2}} 2(z+R)\,dz. \tag{15.7}$$

This integral breaks up into two:

$$I_{r1}' = \frac{1}{2} I_r \Delta\varphi \int_{-\infty}^\infty e^{-\frac{z^2}{\delta_r^2 \ln 2}} 2z dz = 0, \tag{15.8}$$

$$I_{r2}' = I_r \Delta\varphi R \int_{-\infty}^\infty e^{-\frac{z^2}{\delta_r^2 \ln 2}} dz = 2I_r \Delta\varphi R\left(\frac{\sqrt{\pi}}{2}\delta_r \ln 2\right). \tag{15.9}$$

Assuming again $\Delta\varphi = 2\pi$, we get

$$I_r' = 2\pi R I_r \delta_r \sqrt{\pi \ln 2}. \tag{15.10}$$

Substituting for $\ln 2$ its numerical value 0.6931, we obtain from (15.6) and (15.10)

$$\frac{I_r'}{I_c'} = \frac{2\sqrt{\pi}}{\sqrt{\ln 2}}\left(\frac{I_r}{I_c}\right)\left(\frac{R\delta_r}{\delta_c^2}\right) = 2.13\left(\frac{I_r}{I_c}\right)\left(\frac{2\delta_r R}{\delta_c^2}\right). \tag{15.11}$$

In Section 12 we have obtained the following ratio of the half-widths of the profile of the angular distribution of the axial and off-axis SRS radiation for the second and third Stokes components

$$\delta_c = 2\delta_r.$$

Equation (15.11) assumes the simpler form

$$\frac{I_r'}{I_c'} = 2.13\left(\frac{I_r}{I_c}\right)\left(\frac{R}{\delta_c}\right). \tag{15.12}$$

For benzene $R = 3.7 \cdot 10^{-2}$ rad and $\delta_c = 0.9 \cdot 10^{-2}$ rad. We obtain

$$\frac{I_r'}{I_c'} = 8.8\left(\frac{I_r}{I_c}\right). \tag{15.13}$$

If we assume for the ratio $I_r/(I_c + I_r)$ the obtained experimental value 0.04–0.08, then we get $I_r'/I_c' = 0.4$–0.8. This means that the off-axis radiation of the second and higher Stokes components constitutes under our conditions of SRS excitation a noticeable fraction of the total radiation of these components. It is important that this radiation leaves the working channel in which the SRS develops. It therefore causes "losses" which may turn out to be the factor that limits the possibility of conversion of the exciting radiation into higher-order SRS components.

CONCLUSIONS

In our investigations, aimed at studying the angular characteristics of the SRS, we obtain the following principal results.

1. We measured the angular distribution of the first Stokes SRS component in a number of organic liquids (benzene, carbon disulfide, acetone, mixture of acetone with carbon disulfide) and in calcite crystals. A "replica effect" was observed, consisting in the fact that the angular distribution of the first Stokes component approximately repeats the angular distribution of the exciting radiation emerging from the scattering medium.

2. It was observed that self-focusing of the exciting radiation in the scattering medium does not influence the angular distribution of the first Stokes component. The complicated mode composition of the exciting radiation leads to the appearance of a structure and to a broadening corresponding to the distribution of the first Stokes component of the SRS. The replica effect manifests itself in the most distinct form when the SRS is excited by a single-mode laser.

3. We investigated the angular distribution of the paraxial radiation of the second, third, and fourth Stokes components of the SRS in the indicated organic liquid. We have shown that the "replica effect" manifests itself in the angular distribution of the paraxial radiation of all these components, i.e., when the number of the component increases, the width of the angular distribution of the paraxial radiation remains unchanged.

4. The "replica effect" observed in the investigation is explained on the basis of the theory of Stokes – Stokes coherent four-photon SRS processes.

5. We investigated class-II SRS emission in several organic liquids (benzene, carbon disulfide, and nitrobenzene). The rings of the off-axis radiation of the third and fourth SRS Stokes components were studied for the first time ever. The angles and intensities of this radiation were measured as well as those of the radiation of the second Stokes and first anti-Stokes components.

6. A new interpretation is proposed for the off-axis class-II radiation. It differs from the existing one by the fact that it makes no use of the hypothesis that the class-II radiation is excited in thin self-focusing filaments of the exciting radiation in the medium. It is proposed that the class-II radiation is produced as a result of Stokes – anti-Stokes four-photon processes that are subject to the coherence condition. On the basis of this condition, we calculated the angles of the off-axis radiation of the first, second, and third anti-Stokes components and of the second, third, and fourth Stokes components in benzene, carbon disulfide, and nitrobenzene. It is shown that the calculated values of the angles are in satisfactory agreement with the experimental values of the angles for the off-axis class-II radiation.

7. We investigated the dependence of the relative intensity of the off-axis and axial class-II SRS radiation in benzene on the excitation conditions. It is shown that at constant intensity of the laser emission the change of the intensity of the axial radiation of the first Stokes component leads to a change in the ratio of the intensities of the axial and off-axis radiations of the second Stokes component. At low intensity of the first Stokes component, only off-axis class-II radiation of the second Stokes component is produced. With increasing intensity of the first Stokes component, the axial radiation of the second Stokes component appears and begins to gain in intensity. At higher intensities of the first Stokes component, the bulk of the second Stokes component goes over into the axial radiation. The relative intensities of the off-axis and axial radiation of the third and fourth Stokes components behave similarly as functions of the axial radiation of the second Stokes component. The observed regularities are theoretically explained.

8. An estimate is presented of the relative intensity of the off-axis SRS radiation to the total intensity of the radiation of the second and higher Stokes components. It is shown that the intensity of the off-axis radiation of the second and higher Stokes components can amount to 40-80% of the intensity of the axial radiation at these components. Consequently, the off-axis radiation can give rise to considerable "losses" in the conversion of the exciting radiation into higher-order SRS components.

I am deeply grateful to the entire staff of the Optical Laboratory of the Physics Institute of the USSR Academy of Sciences for the opportunity of performing the scientific work in this laboratory and for constant friendly relations, to my scientific director Professor M. M. Sushchenskii for suggesting the interesting topic, to A. V. Kraiskii for helpful discussions, and to V. P. Karpova and Yu. N. Pribylov for help with the experiment and in the discussion of the results.

LITERATURE CITED

1. S. A. Akhmanov and R. V. Khokhlov, Problems of Nonlinear Optics [in Russian], VINITI, Moscow (1964).
2. N. Bloembergen, Nonlinear Optics, Benjamin (1965).
3. N. Bloembergen, Am. J. Phys., 35:989 (1967).
4. H. W. Schrötter, Naturwissenschaften, 54:657 (1967).
5. V. A. Zubov, M. M. Sushchinskii, and I. K. Shuvalov, Usp. Fiz. Nauk, 83:197 (1964).
6. V. A. Zubov, M. M. Sushchinskii, and I. K. Shuvalov, Usp. Fiz. Nauk, 89:49 (1966).
7. M. M. Sushchinskii, Raman Spectra of Molecules and Crystals [in Russian], Nauka, Moscow (1969).
8. V. N. Lugovoi, Introduction to the Theory of Stimulated Raman Scattering [in Russian], Nauka, Moscow (1968).
9. Ya. S. Bobovich and A. V. Bortkevich, Usp. Fiz. Nauk, 103:3 (1971).
10. A. Z. Grasyuk, V. F. Efimkov, I. G. Zubarev, V. I. Mishin, and S. G. Smirnov, Pis'ma Zh. Éksp. Teor. Fiz., 8:474 (1968).
11. E. J. Woodbury and W. K. Ng, Proc. IRE, 50:2367 (1962).
12. R. W. Hellwarth, Phys. Rev., 130:1850 (1963).
13. G. Eckhardt, R. W. Hellwarth, F. J. McClung, S. E. Schwarz, D. Weiner, and E. J. Woodbury, Phys. Rev. Lett., 9:455 (1962); Electron. Des., 11:28 (1963).
14. M. Geller, D. Bortfeld, and W. R. Sooy, Appl. Phys. Lett., 3:36 (1963).
15. R. W. Terhune, Bull. Am. Phys. Soc., 8:359 (1963).
16. B. P. Stoicheff, Internal School of Physics "Enrico Fermi," Course XXXI (1963).
17. R. W. Minck, R. W. Terhune, and W. G. Rado, Appl. Phys. Lett., 3:181 (1963).
18. S. Dumarten, B. Oksengorn, and B. Vodar, C. R. Acad. Sci., 259:4589 (1964).
19. G. Eckhardt, D. P. Bortfeld, and M. Geller, Appl. Phys. Lett., 3:137 (1963).
20. E. Garmire, F. Pandarese, and C. H. Townes, Phys. Rev. Lett., 11:160 (1963).
21. C. H. Townes, International School of Physics "Enrico Fermi," Course XXXI (1963).
22. R. Chiao and B. P. Stoicheff, Phys. Rev. Lett., 12:290 (1964); Bull. Am. Phys. Soc., 9:490 (1964).
23. R. W. Hellwarth, F. J. McClung, W. G. Wagner, and D. Weiner, Bull. Am. Phys. Soc., 9:490 (1964).
24. E. Garmire, Phys. Lett., 17:251 (1965).
25. E. Garmire, Thesis, Massachusetts Institute of Technology (1965).
26. H. J. Teiger and P. E. Tannenwald, Phys. Rev. Lett., 11:419 (1963).
27. P. D. Maker and R. W. Terhune, Phys. Rev.; 137:A801 (1965).
28. E. Garmire, Bull. Am. Phys. Soc., 9:490 (1964).
29. G. Bret, C. R. Acad. Sci., 259:2991 (1964).
30. C. L. Tang and T. F. Deutsch, Phys. Rev., 138A:1 (1965).
31. A. I. Sokolovskaya, E. A. Morozova, and A. D. Kudryavtseva, Zh. Prikl. Spektrosk., 18:122 (1973).
32. E. A. Morozova, Diploma Thesis, Moscow Physicotechnical Institute (1972).
33. B. M. Ataev and V. N. Lugovoi, Pis'ma Zh. Éksp. Teor. Fiz., 7:52 (1968).
34. B. M. Ataev and V. N. Lugovoi, Opt. Spektrosk., 26:1045 (1969).
35. V. N. Lugovoi, Zh. Éksp. Teor. Fiz., 51:931 (1966).
36. B. M. Ataev and V. N. Lugovoi, Fiz. Tverd. Tela (Leningrad), 10:1991 (1968).
37. B. M. Ataev and V. N. Lugovoi, Opt. Spektrosk., 27:700 (1969).
38. B. M. Ataev, Candidate's Dissertation, Physics Institute, Academy of Sciences of the USSR (1969).
39. P. E. Tannenwald, in: Abstracts of Papers of the 3rd All-Union Symposium on Nonlinear Optics [in Russian], Moscow University Press (1967), p. 58.
40. A. K. McQuillan and B. P. Stoicheff, Bull. Am. Phys. Soc., 12(17):60 (1967).
41. G. Rivoire, These de doctorat, Paris (1968).
42. G. Rivoire, J. Phys. (Paris), 28:711 (1967).
43. P. Kelley, Phys. Rev. Lett., 15:1005 (1965).
44. G. Mayer and F. Gires, C. R. Acad. Sci., 258:2039 (1964).
45. S. Kielich, Acta Phys. Pol., 31:639 (1967).
46. S. Kielich, Proc. Phys. Soc. London, 90:847 (1967).
47. G. A. Askar'yan, Pis'ma Zh. Éksp. Teor. Fiz., 4:400 (1966).
48. G. A. Askar'yan, Pis'ma Zh. Éksp. Teor. Fiz., 6:672 (1968).
49. B. Kaspronicz and S. Kielich, Acta Phys. Pol., 31:787 (1967).
50. V. I. Talanov, Pis'ma Zh. Éksp. Teor. Fiz., 2:138 (1965).
51. R. Chiao, E. Garmire, and C. Townes, Phys. Rev. Lett., 13:479 (1964).
52. S. A. Akhmanov, A. P. Sukhorukov, and R. V. Khokhlov, Zh. Éksp. Teor. Fiz., 50:1537 (1966).

53. S. A. Akhmanov, A. P. Sukhorukov, and R. V. Khokhlov, Usp. Fiz. Nauk, 93:19 (1967).
54. A. L. Dyshko, V. N. Lugovoi, and A. M. Prokhorov, Pis'ma Zh. Éksp. Teor. Fiz., 6:655 (1967).
55. V. N. Lugovoi and A. M. Prokhorov, Pis'ma Zh. Éksp. Teor. Fiz., 7:153 (1968); Usp. Fiz. Nauk, 111: 203 (1973).
56. A. L. Dyshko, V. N. Lugovoi, and A. M. Prokhorov, Dokl. Akad. Nauk SSSR, 188:792 (1969).
57. V. V. Korobkin, A. M. Prokhorov, R. V. Serov, and M. Ya. Shchelev, Pis'ma Zh. Éksp. Teor. Fiz., 9:108 (1969).
58. M. M. T. Loy and Y. R. Shen, Phys. Rev. Lett., 22:994 (1969); 25:1333 (1970); Appl. Phys. Lett., 19:285 (1971).
59. E. Garmire, R. Chiao, and C. Townes, Phys. Rev. Lett., 16:347 (1966).
60. P. Lallemand and N. Bloembergen, Phys. Rev. Lett., 15:1010 (1965).
61. Y. Shen and Y. Shaham, Phys. Rev. Lett., 15:1008 (1965).
62. G. A. Askar'yan, Zh. Éksp. Teor. Fiz., 42:1567 (1962).
63. V. I. Talanov, Izv. Vyssh. Uchebn. Zaved. Radiofiz., 7:564 (1964).
64. N. F. Pilipetskii and A. R. Rustamov, Pis'ma Zh. Éksp. Teor. Fiz., 2:88 (1965).
65. G. G. Bret and M. M. Denariez, Appl. Phys. Lett., 8:151 (1966).
66. A. I. Sokolovskaya, A. D. Kudryaetseva, T. P. Zhbanova, and M. M. Sushchinskii, Zh. Éksp. Teor. Fiz., 53:429 (1967).
67. N. Bloembergen, P. Lallemand, and A. Pine, IEEE J. Quantum Electron., QE-2:246 (1966).
68. A. N. Arbatskaya and M. M. Sushchinskii, Preprint FIAN, No. 13 (1969).
69. A. Piekara, IEEE J. Quantum Electron., QE-2:249 (1966).
70. M. Maier and W. Kaiser, IEEE J. Quantum Electron., QE-2:256 (1966).
71. C. A. Sacchi and C. H. Townes, Phys. Rev., 174:439 (1968).
72. C. C. Wang, Phys. Rev. Lett., 16:344 (1966).
73. A. I. Sokolovskaya, A. D. Kurdryavtseva, and M. M. Sushchinskii, Nonlinear Optics, Proceedings of the 2nd All-Union Symposium on Nonlinear Optics [in Russian], Nauka, Novosibirsk (1968), p. 277.
74. A. I. Sokolovskaya, A. D. Kudryavtseva, G. L. Brekhovskikh, and M. M. Sushchinskii, Preprint FIAN, No. 169 (1968).
75. A. D. Kudryavtseva, A. I. Sokolovskaya, and M. M. Sushchinskii, Zh. Éksp. Teor. Fiz., 59:1556 (1970).
76. A. D. Kudryavtseva, A. I. Sokolovskaya, and M. M. Sushchinskii, Kvantovaya Élektron. (Moscow), No. 7, 73 (1972).
77. F. Shimizu, U. Bachmann, and B. P. Stoicheff, IEEE J. Quantum Electron., QE-4:425 (1968).
78. A. Szöke, Bull. Am. Phys. Soc., 9:490 (1964).
79. V. A. Zubov, M. M. Sushchinskii, and I. K. Shuvalov, Zh. Éksp. Teor. Fiz., 47:784 (1964); 48:378 (1965).
80. D. P. Bortfeld, M. Geller, and G. Eckhardt, J. Chem. Phys., 40:1170 (1964).
81. V. A. Zubov, M. M. Sushchinskii, and I. K. Shuvalov, Zh. Prikl. Spektrosk., 3:336 (1965).
82. W. Kaiser, M. Maier, and J. A. Giordmaine, Appl. Phys. Lett., 6:25 (1965).
83. A. I. Sokolovskaya, Tr. Fiz. Inst. Akad. Nauk SSSR, 27:63 (1964).
84. V. A. Zubov, A. V. Kraiskii, K. A. Prokhorov, M. M. Sushchinskii, and I. K. Shuvalov, Preprint FIAN, No. 17 (1968); Zh. Éksp. Teor. Fiz., 55:443 (1968).
85. V. V. Ragul'skii and F. S. Faizullov, Pis'ma Zh. Éksp. Teor. Fiz., 6:887 (1967).
86. A. Z. Grasyuk, I. G. Zubarev, V. I. Mishin, and V. G. Smirnov, Preprint FIAN, No. 14 (1973), Kvantovaya Élektron. (Moscow), No. 5(17), 27 (1973).
87. Y. R. Shen and N. Bloembergen, Phys. Rev., 137:A1787 (1965).
88. D. Linde, M. von der Maier, and W. Kaiser, Phys. Rev., 178:11 (1969).
89. S. K. Potapov and M. A. Kovner, Opt. Spektrosk., 27:939 (1969); M. A. Kovner, S. K. Potapov, and A. R. Kristillov, Abstracts of the 5th All-Union Conference on Nonlinear Optics, Kishinev [in Russian], Moscow University Press (1970), p. 101.
90. D. Buckingham, J. Chem. Phys., 43:25 (1965).
91. A. Shavlov, Usp. Fiz. Nauk, 81:745 (1963).
92. H. Nisinawa and F. Takano, J. Phys. Soc. Jpn., 22:1446 (1967).
93. K. A. Prokhorov and M. M. Sushchinskii, Kratk. Soobshch. Fiz., No. 5:48 (1970).
94. F. Aussenegg and U. Desarno, Phys. Lett., 34A:260 (1971).
95. U. Desarno and G. Nath, Phys. Lett., A30:483 (1969).
96. F. Aussenegg and U. Desarno, Opt. Commun., 2:295 (1970).
97. V. T. Platonenko and R. V. Khokhlov, Zh. Éksp. Teor. Fiz., 46:555, 695 (1964).
98. V. N. Lugovoi, Opt. Spektrosk., 20:996 (1966).
99. V. N. Lugovoi, Zh. Éksp. Teor. Fiz., 48:1216 (1965).
100. V. N. Lugovoi, Opt. Spektrosk., 21:293 (1966).

101. V. N. Lugovoi, Opt. Spektrosk., 21:432 (1966).

102. K. Shimoda, Jpn. J. Appl. Phys., 5:86 (1966).

103. K. Shimoda, Jpn. J. Appl. Phys., 5:615 (1966).

104. V. A. Zubov, A. V. Kraiskii, M. M. Sushchinskii, M. I. Fedyakina, and I. K. Shuvalov, Zh. Éksp. Teor. Fiz., 59:1466 (1970).

105. A. N. Arbatskaya and M. M. Sushchinskii, Proceedings of the 4th All-Union Conference on Nonlinear Optics, Kiev [in Russian], Moscow University Press (1968), p. 26.

106. A. N. Arbatskaya and M. M. Sushchinskii, Preprint FIAN, No. 3 (1969).

107. M. M. Sushchinskii, Preprint FIAN, No. 151 (1967).

108. M. M. Sushchinskii, Kratk. Soobshch. Fiz., Nos. 2 and 3 (1972).

109. A. N. Arbatskaya, K. A. Prokhorov, and M. M. Sushchinskii, Zh. Éksp. Teor. Fiz., 62:872 (1972).

110. A. N. Arbatskaya and M. M. Sushchinskii, Zh. Éksp. Teor. Fiz., 66:1993 (1974).

111. E. B. Aleksandrov, A. M. Bonch-Bruevich, N. N. Kostin, and V. A. Khodovoi, Zh. Éksp. Teor. Fiz., 49:1435 (1965).

112. A. Z. Grasyuk, V. I. Popovichev, V. V. Ragul'skii, and F. S. Faizullov, Kvantovaya Élektron. (Moscow), No. 1, 70 (1971).

113. V. V. Bocharov, M. G. Gangerdt, A. Z. Grasyuk, and I. G. Zubarev, Zh. Éksp. Teor. Fiz., 57:1585 (1969).

114. V. V. Bocharov, M. G. Gangerdt, A. Z. Grasyuk, I. G. Zubarev, V. G. Smirnov, and E. A. Yukov, Preprint FIAN, No. 104 (1969).

115. A. Z. Grasyuk, I. G. Zubarev, V. I. Mishin, and V. G. Smirnov, Preprint FIAN, No. 32 (1973).

116. T. M. Makhviladze and L. A. Shelepin, Preprint FIAN, No. 145 (1971).

117. T. M. Makhviladze and L. A. Shelepin, Zh. Éksp. Teor. Fiz., 62:2066 (1972).

118. T. M. Makhviladze and L. A. Shelepin, Phys. Lett., 39A:409 (1972).

119. T. M. Makhviladze and L. A. Shelepin, Izv. Akad. Nauk SSSR, Ser. Fiz., 37:2190 (1973).

120. G. Placzek, The Rayleigh and Raman Scattering, US AEC Report, UCRL-Trans-526(L) (1962).

121. V. A. Chirkov, V. S. Gorelik, G. V. Peregudov, and M. M. Sushchinskii, Pis'ma Zh. Éksp. Teor. Fiz., 10:416 (1969).

122. V. A. Zubov, P. P. Kircheva, and M. M. Sushchinskii, Kratk. Soobshch. Fiz., No. 1:45 (1971).

123. V. G. Cooper and A. D. May, Appl. Phys. Lett., 7:74 (1965).

INVESTIGATION OF THE FORMATION AND SELF-FOCUSING OF STIMULATED RAMAN SCATTERING OF LIGHT IN CONDENSED MEDIA*

A. D. Kudryavtseva

A detailed investigation is reported of the formation of SRS in the propagation direction of the exciting radiation and in the opposite direction. All the principal characteristics of SRS in these directions were measured. It is shown that the asymmetry of the SRS is due to singularities of the propagation of the SRS in the interior of the medium. The connection between SRS formation and self-focusing of light is investigated. It is observed for the first time that in substances having small Kerr constants self-focusing of SRS of light takes place in the absence of self-focusing of the exciting radiation. It is shown that the SRS self-focusing regime changes when the excitation conditions change and depends on the power converted into SRS.

INTRODUCTION

Stimulated Raman scattering of light (SRS) is one of the most interesting phenomena of nonlinear optics. This is evidenced by the vigorous expansion of theoretical experimental research on this phenomenon soon after its discovery [1]. SRS of light yields information on the peculiarities of molecular structure and on the nonlinear properties of the molecular polarizability when the substance is acted upon by powerful electromagnetic field, and uncovers a possibility of studying new wave interactions that occur in the course of development of SRS in space, interactions that do not occur in ordinary Raman scattering (RS) of light. In contrast to ordinary Raman scattering, SRS has a number of properties typical of lasers (coherence, small divergence, high radiation brightness) [2-5].

Detailed investigations of this phenomenon is of undoubted interest from the point of view of quantum electronics, molecular spectroscopy, and a large number of practical applications. The possibility of obtaining coherent radiation with different pulse durations (down to picosecond pulses, as was subsequently found) in a wide range of wavelengths uncovers prospects for the use of SRS to develop tunable lasers, which are needed for the solution of a number of problems in molecular spectroscopy, nonlinear optics, chemistry, and particularly for the initiation of chemical reactions, holography, etc. From this point of view, interest attaches to a detailed study of the properties of SRS of light in the course of its formation and searches for the possibility of controlling its characteristics, such as brightness, homogeneity of the distribution of the energy in the beam, divergence, and pulse duration and waveform.

This explains the tremendous interest of the scientists in investigations of SRS of light and the rapid subsequent development of this branch of nonlinear optics. Simultaneously with the first experimental investigations, a number of workers have developed a semiclassical [6-10] and a quantum [11, 12] theory of SRS.

In these papers they obtained the basic theoretical laws that describe the formation of SRS of light in media. An expression was obtained that accounted for the increase in the intensity of the first Stokes component and of higher-order components as functions of the excitation energy and of the thickness of the scattering layer. Also obtained were expressions for the gain, whose magnitude was determined mainly by the effective cross section and the width of the ordinary Raman scattering of light. The angular distribution of the SRS intensity was described. According to the theory, for the first Stokes component the distribution of the intensity should be symmetrical in the "forward" and "backward" directions, while the anti-Stokes and higher Stokes components should propagate only along the axis of the exciting radiation or at definite angles to the direction of the incident light.

The first experiments have shown, however, that actually numerous deviations from the theoretical premises are observed. Many studies yielded a much larger SRS gain than predicted by the theory [13, 14]. It was

---

*Dissertation for the degree of Candidate of Physicomathematical Sciences, defended January 20, 1975 at the P. N. Lebedev Physics Institute, Academy of Sciences of the USSR. The scientific adviser was Candidate of Physicomathematical Sciences A. I. Sokolovskaya.

shown that the anomalously high gain takes place at a definite thickness of the scattering layer [15] and does not depend on the mode composition of the exciting radiation [13]. It was observed that despite the existing theoretical concepts, the intensity of the first Stokes component of SRS in the propagation direction of the exciting radiation ("forward") is much larger than in the opposite direction ("backward") [16-18]. Deviations from the theory were observed also in the angular distribution of the anti-Stokes and Stokes components of higher order [5, 19-21]. In a number of substances these components were observed at angles larger than theoretically calculated [19, 20]. The first anti-Stokes component appeared not only at an angle but also along the axis of the exciting radiation [18].

The deviation of the experimental results from the theory found no explanation for some time. Following the discovery, in 1962, of self-focusing of light [22, 23] it became clear that powerful laser radiation produces in media a new nonlinear phenomenon, which can greatly complicate the process of the development of a SRS of light. Even in one of the first theoretical papers [24], devoted to self-focusing of laser radiation in substances, it was indicated that self-focusing of light can be the cause of many discrepancies between the theoretical and experimental results in the investigation of SRS. Soon all the "anomalies" observed in the investigation of SRS of light were attributed, without sufficient justification, to the influence of this phenomenon on the development of the SRS.

Self-focusing of light sets in because of a change of the dielectric constant, and consequently also of the refractive index of the medium under the influence of the powerful laser radiation. This is accompanied by deflection of light towards the highest value of the refractive index (towards the beam axis). It was assumed that if the refraction compensates for the diffraction divergence, then "hot" filaments are produced, along which light propagates with high power density.

This trend was diligently followed in recent years, and the results were new important ideas in the understanding of the self-focusing effect. It was shown theoretically [25, 26] and experimentally [27] that under certain conditions the filaments observed when the integral self-focusing picture is recorded are the wakes of foci that travel at high velocities.

The principal mechanism that leads to self-focusing of light in liquids is the change of the refractive index of the light because of the orientational Kerr effect. An opinion was formed that in substances with large Kerr constant the SRS is produced and propagates in self-focusing filaments, where the power density of the exciting radiation is highest. It was shown in [28, 29] that substances with large Kerr constants have a lower SRS threshold than substances with small Kerr constants. As a result, in many subsequent studies the self-focusing threshold for the exciting radiation was identified with the appearance of the SRS [15, 28].

It was assumed that in substances with large Kerr constants the onset and propagation of SRS in light-conducting filaments leads to deviations of the experimental data from the theory. On the contrary, in substances with low Kerr constants, no self-focusing takes place and the SRS obeys the theoretically derived laws.

At the time that the present research was initiated, however, these opinions have not yet been sufficiently corroborated experimentally. There were practically no experimental studies in which self-focusing was observed simultaneously with investigations of the formation of SRS. The mutual relation between these nonlinear effects was not completely clarified. On the other hand, the formation of SRS in media has hardly been investigated. There was no experimental proof that no self-focusing of light occurs in substances with small Kerr constants. The value of the available experimental results was very low because of insufficiently developed measurement procedures.

The purpose of the present study was to investigate the formation of SRS in a medium in which other nonlinear effects (self-focusing of the exciting radiation, SMBS) do not play a significant role. Such a substance is liquid nitrogen. It was proposed to carry out subsequently analogous investigations in substances with different values of the Kerr constant (for different self-focusing ability).

It was proposed to investigate the mutual relation between self-focusing and SRS of light in substances with different Kerr constant under various excitation conditions.

It was also of interest to investigate the influence of the temperature of the scattering medium on the SRS formation. The change of temperature causes a change in the optical constants of the medium, of the Kerr constant, and of the probability of Raman scattering of light. Thus, by varying the temperature, it is possible to vary the regime of the self-focusing of light in the medium and the conditions for the excitation of SRS of light and to investigate the interaction of these effects.

The formation of SRS in the excitation direction and in the opposite direction has hardly been investigated prior to the start of the present study. Only the ratio of the total energies in these directions was measured.

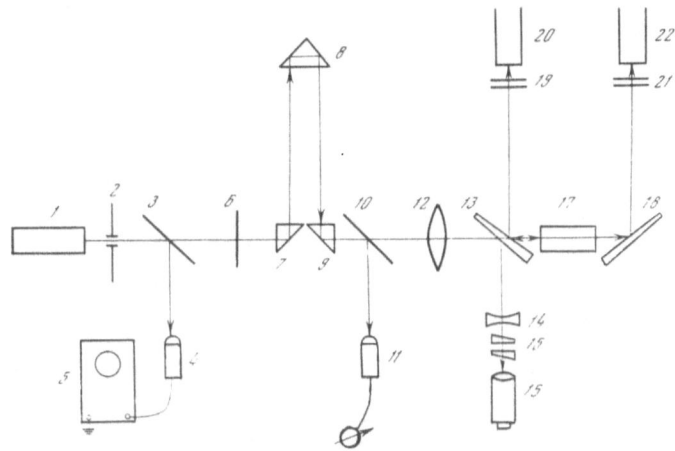

Fig. 1. Diagram of experimental setup. (1) Ruby laser;
(2) diaphragm; (3, 10, 13, 18) turning plates; (4) photodiode; (5) S1-4 oscilloscope; (6, 19, 21) light filters; (7, 8,
9) turning prisms; (11) Kozyrev thermocouple connected
to an M17-3 galvanometer; (12) lens focusing the radiation
into the medium; (14) scattering lens; (15) Fabry–Perot
etalon; (16) camera with F = 840 mm; (17) cell with medium; (20, 22) receivers of SRS of light.

We planned in the present study to measure in the forward and backward direction not only the energy but also the divergence of the radiation, the pulse duration, the line width, the intensity distribution in the near and far fields, and the SRS power per unit solid angle, as well as to investigate the dependence of these quantities on the excitation conditions (on the thickness of the scattering layer, on the energy of the exciting radiation).

Much attention was paid to the development of the measurement procedure. The measurements were made photographically and by photoelectric means. Under fixed experimental conditions we measured simultaneously all the principal parameters of the SRS. The mode composition and stability of the laser radiation were monitored in the course of the measurements.

CHAPTER I

EXPERIMENTAL SETUP AND MEASUREMENT PROCEDURE

1. Experimental Setup

In the present study, the SRS was excited by a giant Q-switched ruby-laser pulse (λ = 6943 Å). We used a ruby rod with sapphire end pieces 8 mm thick, 180 mm long (length of active part 120 mm). The laser operated in a single-spike regime. In each experiment we checked the energy, pulse duration, and the mode composition of the laser radiation. In the exciting light spectrum we observed a single line 0.015 cm^{-1} wide. The divergence of the laser radiation was 3.5'. The energy distribution and the beam cross section in the far field were close to Gaussian. The observed inhomogeneities in the distribution of the energy in the beam of the exciting radiation were small. The small beam divergence and the proximity of the energy distribution to Gaussian form are evidence that the main contribution to the ruby emission was made by the TEM$_{00}$ mode, while the contributions of the remaining transverse modes were small. The exciting-radiation pulse was smooth, single, with duration 20 nsec. The maximum laser power was 10 MW.

The optical system of the experimental setup for the investigation of the SRS formation in the propagation direction of the exciting radiation and in the opposite direction is shown in Fig. 1.

The radiation of the ruby laser 1 passed through diaphragm 2. The number of spikes in the pulse of the exciting radiation was monitored by photodiode 4, to which part of the radiation was diverted by glass plate 3. The signal from the photodiode was fed to the S1-4 oscilloscope. The energy of the exciting radiation could be varied by placing in the beam calibrated neutral light filters 6. To prevent feedback between the medium and the laser, an optical delay line 4 m long was used. To delay the line we used a system of turning prisms 7-9.

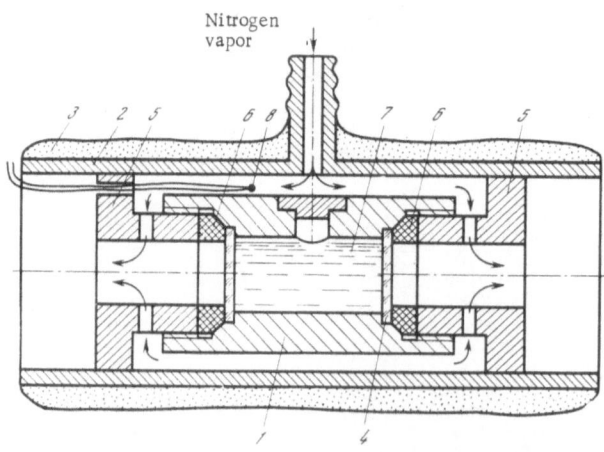

Fig. 2. Cell for the investigation of SRS of light in liquids at various temperatures. (1) Body of the cell; (2) metallic jacket; (3) asbestos cover; (4) cell window; (5) plugs; (6) Teflon liners; (7) investigated liquid; (8) thermocouple.

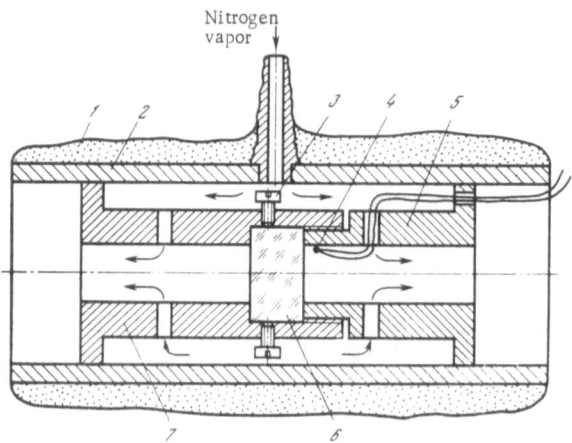

Fig. 3. Cell for the investigation of SRS of light in crystals at various temperatures. (1) Asbestos cover; (2) jacket; (3) screw that secures the crystal in the cell; (4) thermocouple; (5, 7) parts of the cell body; (6) crystal.

To measure the energy of the exciting radiation, part of the light was diverted by plate 10 to Kozyrev receiver 11, which was graduated with the aid of a KI-1 calorimeter placed in the position of the cell with the scattering medium. The spectral composition of the exciting radiation was monitored with the aid of Fabry–Perot etalon 15 and camera 16. A diverging lens 14 was placed in front of the etalon to obtain more uniform illumination. The exciting radiation was focused into the investigated medium 17 by lens 12 with focal length 250 mm.

To eliminate the lasing due to reflection from the end faces of the sample, the windows of the cell or the end faces of the crystal were mounted with an angle 10° to each other. In those cases when the windows of the cell were parallel, the cell was mounted at a small angle to the optical axis of the system, so that the light reflected from the windows did not strike the radiation receivers.

The setup made it possible to register simultaneously SRS in the excitation direction and in the opposite direction. To this end, the radiation leaving the investigated sample was reflected by two glass wedges 13 and 18 to receivers 20 and 22. Fabry–Perot etalons, photographic plates, or other receivers were placed in the positions 20 and 22, depending on the measurements performed.

40

The required SRS components were separated by a selective glass or interference light filters 19 and 21, which were placed ahead of the receivers 20 and 22.

In some experiments, the SRS was registered with the aid of a spectrograph. In this case turning prisms of special construction were used to make the light beams travel at different heights in the forward and backward directions, and to direct them to the spectrograph slit.

To investigate the SRS of light at various temperatures of the scattering medium, the cells with the liquid and with the crystals were placed in special thermostatically controlled jackets (Figs. 2 and 3). The heating was with electric current flowing through a coil wound around the jacket. Cooling was produced by blowing cold liquid-nitrogen vapor in the space between the jacket and the cell. To prevent the windows of the cell from fogging, sleeves were placed on the ends of the cell in the course of cooling, as a result of which the nitrogen vapor flowed over the cell windows. The temperature of the medium was measured with a graduated copper—constantan thermocouple connected to an M-95 galvanometer.

To investigate SRS in liquid nitrogen and in calcite cooled to liquid-nitrogen temperature, dewars with sealed-in windows were used. The calcite crystal was placed in this case in the dewar on a copper holder in such a way that the liquid nitrogen was in contact with the lower face of the crystal.

2. Procedure for Measuring the Parameters of the SRS of Light

In the present study we measured the following: energy, pulse duration, divergence, and the spatial distributions of the SRS of the light in the excitation direction and in the opposite direction.

Depending on the purpose of the measurements, we used either a photographic or a photoelectric registration method. In the photoelectric registration of the SRS we used Kozyrev thermocouples, FÉK-15 coaxial photocells, or an ÉLU-FT photomultiplier. When the SRS energy was high enough, we measured it with a Kozyrev radiation thermocouple, the signal from which was fed to a mirror galvanometer M 17/3. The scale of the galvanometer was calibrated against the KI-1 calorimeter. The sensitivity of the calorimeter, calibrated against an F116/1 galvanometer, was 0.033 J/μA. The sensitivity of the thermocouples was 0.7-0.8 V/W and was constant in the investigated range of wavelengths. The light rays propagating in the excitation direction and in the opposite direction were focused by cylindrical lenses on the receiving areas of the thermocouples, which measured 2 × 12 mm. The receivers were aligned with the aid of an LG-55 He—Ne laser. In the course of the alignment the cell was replaced by a semitransparent mirror perpendicular to the beam of the exciting radiation and the receivers were arranged in a way such that they were stricken by the light passing through the mirror and reflected from the mirror. To eliminate systematic errors from the measurements of the ratio of the forward and backward SRS energies, due to slight differences between the sensitivities of the receivers, the receiving units were periodically interchanged.

From measurements in a wide energy interval (from 10^{-10} J), we used a receiver having a high sensitivity, namely an ÉLU-FT photomultiplier. It had seven amplification stages. Its gain was $2.4 \cdot 10^6$, and the integral sensitivity 21.6 A/lum. The signal from the ÉLU-FT was applied to an I2-7 oscilloscope, and the pulse on the oscilloscope screen was photographed with RF-3 film (sensitivity 1100 reciprocal roentgens). The receiver was graduated against signals of known energy.

The graduation made it possible to determine the ratio of the energy values and of the areas under the pulse on the oscillogram.

In a number of cases, the measurements were made photographically with a spectrograph. This made it possible to separate simultaneously several SRS components and to measure their relative intensities. We use a "Huet B-III" three-prism spectrograph with a dispersion 47 Å/mm in the red region. When the SRS was registered in the forward and backward directions, it was necessary to verify experimentally that the optical paths of the light propagating in these directions were completely identical. To this end, the cell with the medium was replaced by a plane-parallel plate and light from the gas laser was directed first in the direction of propagation of the exciting radiation and then in the opposite direction. The intensities of the light beams reflected from the plate in the forward and backward directions were compared. It turned out that these intensities were equal within the limits of the measurements errors.

For the photographic measurements we used the photographic plates infra-760, infra-840, I-750, I-850, I-1050, and Rot Rapid. The photographic density was measured with an MF-4 microphotometer and was converted into intensity with the aid of a photographic-density curve. The latter was plotted using density markers photographed with a stepped attenuator illuminated by the investigated radiation. To ensure that the attenuator was uniformly illuminated over its height, a scattering lens and a ground glass were placed in front of the attenuator.

The spatial distribution of the SRS was also measured photographically. To investigate the energy distribution over the beam cross section, the photographic plate was placed directly in the beam of the investigated radiation at a distance of approximately one meter from the scattering medium. The beam divergences were determined by measuring the pattern of the angular distribution of the SRS of the light in the focal plane of the camera. In this case the divergence was determined from the relation $\theta = d/F$, where d is the half-width of the spot and F is the focal length of the camera.

To measure the duration of the SRS pulses and of the exciting light, the receiver employed was an FÉK-15 coaxial photocell or an ÉLU-FT photomultiplier. Both receivers have a maximum sensitivity in the 8000-8500 Å region. The resolution time of the FÉK-15 photocell is $3 \cdot 10^{-10}$ sec, and its sensitivity is $27 \, \mu A/\text{lum}$. At higher sensitivity, the time resolution of the ÉLU-FT photomultiplier is $5 \cdot 10^{-9}$ sec. The signal from the receiver was fed to an I2-7 oscilloscope. The pulse and the time markers were photographed from the screen of the oscilloscope with a camera equipped with a $1:2$ "Jupiter-9" lens. The pulse duration was measured on the oscillogram at half the maximum height.

To determine the excitation-power density, the total energy of the exciting radiation was divided by the duration of the pulse (measured at half-height of the pulse) and by the area of the beam cross section in the caustic region.

CHAPTER II

FORMATION OF SRS OF LIGHT IN THE PROPAGATION DIRECTION OF THE EXCITING RADIATION AND IN THE OPPOSITE DIRECTION

1. Survey of the Literature

Raman scattering of light is a process wherein a molecule goes over into a new state, and the scattered light photon has a frequency that differs from that of the incident radiation by an amount equal to the molecular vibration.

The frequency of the scattered photon is

$$\omega = \omega_0 \pm \omega_{mn}, \tag{1}$$

where ω_0 is the frequency of the incident photon and ω_{mn} is the molecule-transition frequency.

The total probability of the scattering of a photon $\hbar\omega_0$ with transition of the molecule from the state m into the state n is [30]

$$W_{mn} = \frac{(2\pi)^3}{\hbar^2} \sum_{jj'} \int \rho_j(\omega_0 \Omega) \left[\frac{\hbar\omega^3}{(2\pi c)^3} + \rho'_{j'}(\omega, \Omega') \right] |S_{mn}^{mj, m'j'}| \, d\omega \, d\Omega \, d\Omega', \tag{2}$$

where

$$S_{mn} = \frac{1}{\hbar} \sum_r \frac{(e\mathbf{p}_{mr})(e'\mathbf{p}_{rn})}{\omega_{rm} - \omega_0} + \frac{(e\mathbf{p}_{mr})(e'\mathbf{p}_{rn})}{\omega_{rn} + \omega_0}, \tag{3}$$

$\rho_j(\omega_0, \Omega)$ is the energy density of the scattered radiation per unit volume, per unit frequency ω_0, and per unit solid angle Ω; $\rho'_{j'}(\omega, \Omega)$ is the radiation-energy density at the frequency ω; the index j = 1, 2 corresponds to the two possible polarizations; e is a unit vector of the polarization of the photon to be scattered; e' is the same for the scattered photon; $\mathbf{p}_{mn}$ is the matrix element of the dipole moment of the molecule, corresponding to the transition m → n.

The total energy scattered per unit time is

$$P_{mn} = \hbar\omega W_{mn}. \tag{4}$$

At low intensities of the radiation interacting with the molecule, the energy density $\rho'(\omega, \Omega')$ of the scattered radiation is much less than $\hbar\omega^3/(2\pi c)^3$. We can then neglect the second term in the square brackets of Eq. (2), which will then describe the ordinary Raman scattering of light. If the second term is large enough and cannot be neglected, then an important role is assumed by stimulated processes. The probability W_{mn} is then increased if the radiation already contains the frequency $\omega = \omega_0 + \omega_{mn}$, which can appear as a result of of ordinary Raman scattering of light.

Thus, the analysis of ordinary Raman scattering has already yielded a formula that includes the stimulated-scattering process. As a rule, however, the term describing this process was neglected, since the energies realistically attained at that time were too low to permit experimental observation of the stimulated scattering.

With the advent of lasers, the situation changed. The high intensities of the laser radiation made it possible to observe not only the ordinary but also the stimulated Raman scattering of light. There are several different approaches to the theoretical analysis of the SRS phenomenon; they describe different aspects of this process.

Many singularities of SRS can be understood on the basis of the semiclassical theory [6-10]. In this case one considers the interaction of the molecule with the electric field, the latter constituting an assembly of a certain number of plane waves that differ in frequency by an amount ω_{mn}. This method permits calculation of the exciting-radiation power converted into Stokes and anti-Stokes waves and to find the conditions under which amplification of these waves takes place (the phase-synchronism conditions), and consequently, to find also the angles at which the different SRS components propagate.

One of the methods of theoretically analyzing the SRS is to describe it as a nonlinear-optics phenomenon [7, 31, 32]. At high field intensities the polarization is connected with the field intensity by the relation

$$\mathscr{P} = \varkappa\mathbf{E} + \mathscr{P}^{\text{nonlin}} = \mathscr{P}^{\text{lin}} + \mathscr{P}^{\text{nonlin}} , \qquad (5)$$

where $\varkappa$ is the linear susceptibility, $\mathscr{P}^{\text{lin}}$ is the linear polarization, and $\mathscr{P}^{\text{nonlin}}$ describes the deviation from linearity.

The Fourier component of the polarization, corresponding to the frequency ω_p, is

$$\mathscr{P}_i^{\text{nonlin}}(\omega_p) = \chi_{ijk}(\omega_p, \omega_q, \omega_r) E_j(\omega_q) E_k(\omega_r) + \chi_{ijkl}(\omega_p, \omega_q, \omega_r, \omega_s) E_k^*(\omega_r) E_l(\omega_s) E_j(\omega_q) + \dots \qquad (6)$$

The nonlinear susceptibilities of various orders describe the nonlinear properties of the medium. In the approach, the Rayleigh and Raman scattering are described by susceptibilities in the form of tensors of the third rank.

Calculation of the vibrations of the molecule in the light field, with account taken of the nonlinearity of the medium, yields the following equation for the increase of the Stokes-wave amplitude:

$$\frac{dE_s}{dz} = + \frac{2\pi\omega_s^2}{k_{sz}c^2} |\chi_s''| |E_0|^2 |E_s|, \qquad (7)$$

where χ_s'' is the imaginary part of the nonlinear susceptibility for the Stokes wave and z is the longitudinal coordinate.

At the same time, the decrease of the amplitude of the initial wave is described by the equation

$$\frac{dE_0}{dz} = - \frac{2\pi\omega_0^2}{k_{0z}c^2} |\chi_s''| |E_s|^2 |E_0|. \qquad (8)$$

A solution of this system was obtained by Loudon [7], and also by Platonenko and Khokhlov [10]. From these equations it is easy to change over to the equations for the intensities of the SRS and of the exciting radiation, or for the photon numbers. Similar expressions can be obtained also in the quantum-mechanical approach [12]. The solution of the system yields an exponential dependence of the intensity of the SRS on the intensity of the exciting radiation. This dependence can be written in the form [32, 33]

$$I_s(l) = I_s(0) e^{gII_0}, \qquad (9)$$

where $I_s(l)$ is the SRS intensity; $I_s(0)$ is the intensity of the ordinary Raman scattering; I_0 is the intensity of the exciting radiation; l is the thickness of the scattering layer; g is the gain.

The region in which this solution is valid is limited. The solution does not take into account the conversion of the energy into SRS components other than the first Stokes component, and does not describe the saturation of the first SRS Stokes component.

An expression that describes more completely the dependence of the SRS power on the excited-radiation power was obtained in [4, 34]. For the case of small absorption, the power of the first SRS Stokes component is

$$P_s = \frac{(\omega/\omega_1)(n_1/n)^3 b \left[e^{a(P_0+b)} - 1\right]}{1 + (b/P_0) e^{a(P_0+b)}} , \qquad (10)$$

where P_S is the power of the first Stokes component of the SRS; ω, ω_0 are the frequencies of the first Stokes SRS component and of the exciting radiation, respectively; n and n_0 are the refractive indices for the first Stokes component of the SRS and for the exciting radiation; P_0 is the power of the exciting radiation;

$$a = \frac{4\pi\sigma c N_0 n_0^2 l}{\hbar\omega^3\lambda S}; \qquad \upsilon = \frac{\hbar\omega_0\omega^2\delta\Omega S\,(n/n_0)^3}{4\pi c n_0^2},\qquad (11)$$

where σ is the absolute cross section of the ordinary Raman scattering; N_0 is the total number of molecules per unit volume; δ is the line width in cm^{-1}; Ω is the solid angle in which the radiation propagates in vacuum; $\hbar$ is Planck's constant; S is the cross section of the beam in the vacuum.

This formula describes the region of the transition from the ordinary Raman scattering to SRS, the region where the SRS power has an exponential dependence on the exciting-radiation power, and the region of saturation of the SRS. When plotted logarithmically, the exponential region is represented by a straight line whose slope characterizes the SRS gain.

An expression for the SRS gain was obtained by a number of authors. According to [35] the gain can be expressed in the form

$$gI_0 = (4\pi\Gamma/\sqrt{\varepsilon})k_s\,|\,E_0\,|^2;\qquad (12)$$

$$\Gamma = \frac{3\lambda^4 N Q_0}{2^8\pi^5\hbar\delta},\qquad (13)$$

where k_S is the wave vector of the Stokes scattering; $|\,E_0\,|^2$ is the square of the amplitude of the field of the exciting radiation; ε is the dielectric constant of the medium; λ is the wavelength of the light; N is the number of molecules per unit volume; Q_0 is the Raman-scattering cross section per molecule; $\hbar$ is Planck's constant; δ is the width of the Raman-scattering line.

The gain should thus be independent of the direction of propagation of the SRS. It follows also from the Placzek formula that the SRS gain is independent of the direction. In [35] was calculated the SRS intensity for a rod of length l with cross-sectional area S, i.e., for a geometry close to the experimental conditions under which SRS is observed. It was shown that the main contribution to the radiation is made by the priming sources located in the vicinity of the rod end faces. The intensity of the scattered radiation, as a function of the angle ψ between the rod axis and the observation direction, has two equal maxima (in the forward and backward directions), each with width $\Delta\psi \approx \sqrt{S}/l$. If the rod is long enough ($l/\sqrt{S} \gg 1$), then the radiation should have high directivity ($\Delta\psi \ll 1$). This conclusion is valid in a region far from the saturation of the SRS. It was observed even in the first experimental studies of SRS [16-18] that the first Stokes component of the SRS propagates in the form of a narrow beam (within 5°) in the direction of propagation of the exciting radiation and in the opposite direction. It was noted there that the SRS energies are different in these two directions. Thus, Stoicheff [16] has found that the energy of the first Stokes SRS component in the forward directions in liquids is larger by approximately one order of magnitude than in the backward direction. Maker and Terhune [17] have reported that the SRS energy in benzene is different in the excitation direction from that in the opposite direction. For SRS in benzene, with oscillations at the frequency 992 cm^{-1}, the ratio of the energies in the forward and backward directions was 2:1 for the first and second Stokes components; SRS for the oscillation with frequency 3064 cm^{-1} was observed only in the forward direction.

Bret and Denariez [18] have attempted to connect the asymmetry of the SRS energy with self-focusing of the exciting radiation. They measured the ratio of the energies of the first Stokes SRS component in the forward and backward directions in substances having different Kerr constants. It was found that in acetone, a substance with a small Kerr constant, this ratio is equal to unity (at an exciting-radiation power from 10 to 20 MW), whereas in substances with large Kerr constant it is larger than unity.

Thus, at the time when we started the present research (1966), there were only few reported measurements of the ratio of the total SRS energies in the forward and backward directions. The investigations were carried out, as a rule, in substances in which intense stimulated Mandelstam–Brillouin scattering (SMBS) and self-focusing of the exciting radiation were produced simultaneously with the SRS of the light. The interaction of the different nonlinear effects complicated the picture of the development of the SRS of the light. The results obtained by various authors were frequently not in agreement, since the experiments were performed in essentially different conditions. The characteristics of the exciting radiation were not known. In the first experiments, the Q-switch of the laser was usually a rotating prism. The emission of such a laser is usually not single-mode, and the pulse frequently consisted of several spikes, the number and intensity ratio of which changed from flash to flash. In addition, in the first studies they did not eliminate the feedback between the laser and the cell with the investigated substance.

In the course of our investigations we observed that the asymmetry of the SRS in the forward and backward directions is not confined to the energy ratio. Accordingly, we undertook in the present study a detailed investigation of the process of formation of the SRS of the light in the direction of propagation of the exciting radiation and in the opposite direction. We measured the energy and duration of the pulse, the divergence of the beam and the energy distribution in its cross section, as well as the spectral width of the line at various excitation-power densities and scattering-layer thicknesses.

The principal object of the investigation of the formation of SRS of light in the forward and backward direction was liquid nitrogen. In liquid nitrogen the SMBS threshold is much higher than the SRS threshold, as was verified in experiment. The Kerr constant in liquid nitrogen is small, so that self-focusing of the exciting light could not play a significant role. Similar investigations were made on a number of other substances.

The measurements were performed with a laser that generated practically one longitudinal mode. The laser-emission stability was monitored in each experiment.

2. Ratio of SRS Energies in the Forward and Backward Directions under Various Excitation Conditions

In the SRS spectrum of liquid nitrogen, in the direction of the propagation of the exciting radiation, we observed the first and second Stokes and the first and second anti-Stokes components of the SRS. In the backward direction we could register only the first Stokes component. The SMBS, the first anti-Stokes components, and the higher components were not observed in the backward direction at the excitation energies and scattering-layer thicknesses employed by us. The ratio of the energies of the first Stokes component of the SRS in liquid nitrogen in the forward and backward directions was measured in scattering layers 8 to 100 mm thick at a constant power density of the exciting radiation (1800 MW/cm^2) [36]. The results are listed in Table 1, where E_f / E_b is the ratio of the SRS energies in the forward and backward directions, and l is the thickness of the scattering layer.

As seen from the table, in a vessel 8 mm thick the ratio of the SRS forward and backward energies is close to unity. In vessels of larger thickness, the SRS energy in the forward direction is larger than backwards.

The change of the power density of the exciting radiation at a constant thickness of the scattering layer also changes the ratio of the energies of the first Stokes SRS components in the forward and backward directions. The value of this ratio at an exciting-radiation power density ranging from 600 to 6000 MW/cm^2 was measured in a liquid-nitrogen layer of thickness 20 mm. Figure 4 shows plots of the energies of the first Stokes SRS component in liquid nitrogen in the forward and backward directions, in relative units, as well as a plot of the ratio of the energies in these directions against the power density of the exciting radiation. It is seen that the energies of the forward and backward first Stokes components of the SRS do not reach the saturation region simultaneously. In the backward direction saturation is reached at much higher densities of the exciting radiation than forward. The energy ratio changes correspondingly, and reaches unity when the forward and backward scattering reach saturation.

In the present study we investigated the ratio of the energies of the SRS in the forward and backward directions in carbon disulfide and in calcite at a constant exciting-radiation energy and at various temperatures of the scattering medium [37-40]. In contrast to liquid nitrogen, there is intense SMBS in these substances, and in carbon disulfide, in addition, self-focusing of the exciting radiation is observed. It is known that when the temperature of the scattering medium is changed the ratio of the contribution of the various nonlinear effects also changes. When the medium is cooled the SMBS intensity decreases while the SRS intensity increases because of the increased Raman-scattering probability [41,42], and the self-focusing of the light can change as a result of the increase in the Kerr constant. The investigations of the forward/backward SRS energy ratios at various temperatures of the scattering medium are therefore of definite interest.

The temperature of the calcite was varied in the range from +150 to −196°C, and that of carbon disulfide from +20 to −90°C. The results of the measurements of the ratio of the energies of the SRS components in the forward and backward directions at various temperatures, for carbon disulfide and calcite, are shown in Tables 2 and 3, respectively. The measurements were made at an exciting-radiation power density 1-2 GW/cm^2 in layers of thickness 30 mm (CS$_2$) and 10 mm (calcite).

It should be noted that, in contrast to the liquid nitrogen, considerable instability of the SRS energies in the foreward and backward directions was observed in carbon disulfide and in calcite. The results listed in

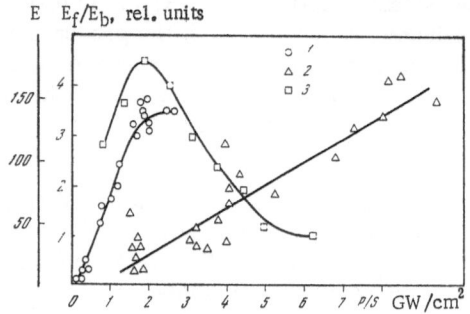

Fig. 4. Plots of the energies of the first Stokes component of SRS in the forward (1) and backward (2) directions and of the energy ratio in these directions (3) vs the exciting-radiation power density.

TABLE 1

l, mm	8	20	50	100
E_f/E_b	1.6 ± 0.6	4.8 ± 1	3.0 ± 0.5	5 ± 0.6

Tables 2 and 3 are averages of 15-30 measurements. The arithmetic mean errors in the calcite were 20-30%. In carbon disulfide, where the instability was even larger, the deviation from the mean value reached sometimes 50%.

For both substances, in contrast to liquid nitrogen, the Stokes and anti-Stokes SRS components were observed in both the forward and in the backward directions. As seen from the tables, the SRS energy in all the cases was larger in the propagation direction of the exciting radiation than in the opposite direction.

The instability of the forward and backward SRS at a given medium temperature, and the change of the ratio E_f/E_b at different temperatures, indicate that the effects that compete with the SRS can influence substantially the ratio of the energies in the excitation direction and in the opposite direction. It is difficult to establish definite relations for the E_f/E_b ratio in this case, since this calls for a detailed analysis of the development and simultaneous interaction of several nonlinearity effects.

3. Investigation of the Duration and Waveform of the Pulses
in the Forward and the Backward Directions in Liquid Nitrogen

We investigated the waveform and duration of the pulses of the exciting radiation before and after passage through the medium, and of the pulses of the first SRS Stokes component in the forward and backward directions in layers of liquid nitrogen of various thicknesses [43]. The exciting-radiation pulse had a smooth form prior to passing through the medium, and its duration was 20 nsec. Oscillograms of the pulses of the exciting radiation after passing through the medium and of the first Stokes component of the SRS in the forward and backward directions in liquid-nitrogen layers 20 and 100 mm thick are shown in Fig. 5. The duration of the SRS pulse in the forward direction was longer under our conditions than the duration of the exciting radiation, and when the thickness of the scattering layer was increased from 20 to 100 mm, it changed from 25 to 38 nsec. The increase of the thickness of the scattering layer led also to a change in the waveform of the SRS pulses and of the exciting-radiation pulses passing through the medium. Dips appeared on the pulses and were

TABLE 2

t, °C	Component			
	1St[*]	2St	1aSt[*]	2aSt
+150	5.0 ± 2.3	6.7 ± 4.3	69 ± 25	—
+100	(2) †	2.5 ± 0.6	108 ± 39	—
+ 20	1.8 ± 0.8	2.3 ± 1.5	11.5 ± 5.2	2.1 ± 1.7
— 20	3.0 ± 1.5	3.1 ± 2.5	21 ± 16	4.2 ± 1.8
— 60	10.4 ± 5	1.2 ± 0.3	15.3 ± 7	2.4 ± 1.7
—110	5.9 ± 2.4	8.0 ± 5	4.2 ± 2.8	(1.4) †
—196	3.6 ± 0.6	1.4 ± 1.1	82 ± 44	4.4 ± 3

[*]St — Stokes, aSt — anti-Stokes.
†The result obtained from one measurement.

TABLE 3

t, °C	Component		
	1 St	2 St	1 aSt
+20	12±5	4.3±2.7	1.6±0.5
—40	8±3	8.2±4	2.9±0.6
- 96	10±3	160±40	6.7±1.7

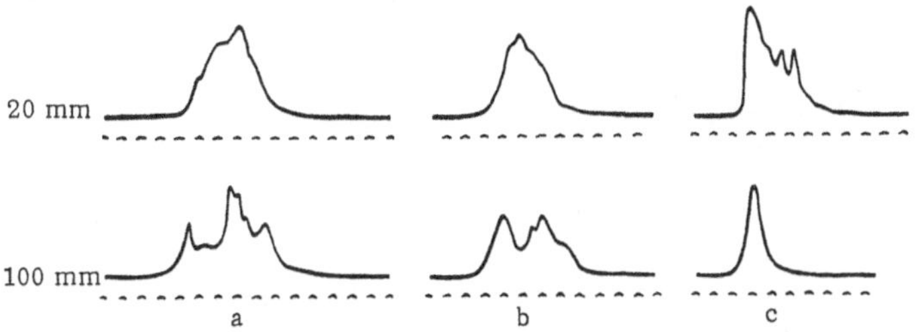

Fig. 5. Oscillograms of pulses of exciting radiation after passage through the medium (a) and of the pulses of the first Stokes SRS component in the forward (b) and backward (c) directions in layers of liquid nitrogen 20 and 100 mm thick. The distance between markers is 10 nsec.

due to the conversion of the energy from the exciting radiation into the first Stokes component and from the first Stokes component into higher-order components. The characteristic features of the pulses, such as the number of groups, the slope of the front, and the width (accurate to 10-15%) remained the same from flash to flash.

In the backward direction, the SRS pulse had a steeper leading front. With increasing thickness of the scattering layer from 20 to 100 mm, the pulse duration decreased from 25 to 9 nsec. The reproducibility of this result was very good, and the pulse did not change in duration and maintained a smooth waveform from flash to flash.

4. Spatial Asymmetry of SRS in Liquid Nitrogen

The energy distribution in the beam cross section and the radiation divergence of the first Stokes SRS component in the forward and backward directions were investigated in layers of liquid nitrogen of various thicknesses and various excitation-power densities [36, 44]. The parameters of the SRS in these directions are closest when the scattering layer is thin (up to 20 mm). The first Stokes SRS component is radiated in this case in the form of a rather broad diffuse beam. With increasing thickness of the layer of the scattering medium, the parameters of the radiation in the forward direction changed little, whereas an appreciable spatial narrowing is observed for the backward beam. The change of the structure of the field and the cross section of the beam of the first Stokes SRS component in the backward direction with changing thickness of the scattering layer is illustrated in Fig. 6. It shows photographs of the SRS field and the results of their photometry. The beam cross sections were photographed at a distance 300 mm from the cell windows. As seen from the figure, at a liquid-nitrogen layer thickness 8 mm the beam cross section constitutes a diffuse spot. With increasing thickness of the scattering layer radiation with greater brightness and much smaller divergence is formed in the center of the diffuse spot. At a liquid-nitrogen layer thickness 20 mm, intense scattering appears in the center of the beam of the first Stokes SRS component. At a scattering-layer thickness 100 mm, practically the entire energy of the first Stokes SRS component is concentrated in a narrow beam that greatly exceeds in brightness the diffuse part of the scattering. At layer thicknesses 50 and 100 mm, a rather intense ring is observed around the bright spot.

The cross section of the first Stokes component in the backward direction was of this form only at an exciting-radiation power density exceeding 1.25 GW/cm^2. Our experiments have shown that the structure of the SRS beam in the backward direction changed with decreasing exciting-radiation power density [43]. At an excitation power density less than 1.5 GW/cm^2, the bright spot broke up into several spots which became more and more blurred with further decrease of the power density. The exciting-radiation power was varied in this case by means of neutral light filters; experiment has shown that the pattern of the energy distribution in the field of the exciting radiation was not distorted.

47

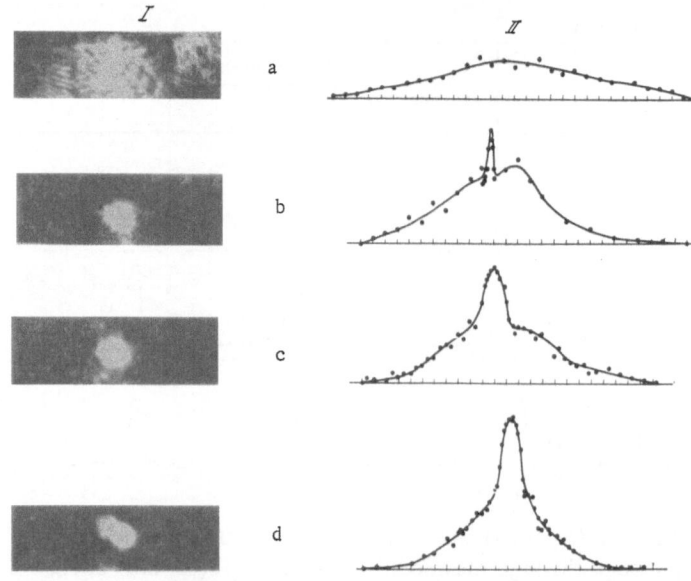

Fig. 6. Distribution of the energy in the cross section of the beams of the first Stokes SRS component in the backward direction at various thicknesses of the liquid-nitrogen layer. (a) 8 mm; (b) 20 mm; (c) 50 mm; (d) 100 mm. (I) Photograph; (II) results of photometry. The exciting-beam power density is 2 GW/cm^2.

The cross section of the beam of the first Stokes component in the forward direction changed little with changing thickness of the scattering layer. It constituted a diffuse spot surrounded by an absorption ring corresponding to the conversion of the energy into radiation of the first anti-Stokes SRS component. This ring is particularly noticeable at a liquid-nitrogen layer thickness 50 or 100 mm. The intensity at the maximum of the beam of the first Stokes component hardly increased with increasing thickness of the scattering layer; only the intensity at the edges of the beam increased.

Figure 7 shows the results of the photometry of the cross sections of the beams of the first Stokes SRS component in liquid nitrogen in the forward and backward directions, obtained on photographic plates located at equal distances from the dewar. The thickness of the scattering layer was 100 mm.

The divergence of the radiation of the first Stokes SRS component in liquid nitrogen was measured in the forward and backward directions in scattering layers of various thicknesses (from 8 to 100 mm) [43, 44]. The divergence of the exciting radiation behind the focus of the lens was 46' ± 2' (in the absence of the medium). After passing through the liquid nitrogen, the divergence of the exciting radiation increased to 102' ± 10'. Table 4 shows the divergence of the beams of the first Stokes SRS component in the forward and backward directions in liquid-nitrogen layers 8, 20, 50, and 100 mm thick. As seen from the table, at a small thickness of the scattering layer, the divergences of the beams of the first SRS Stokes component are close in the forward and backward directions, whereas at a large thickness the divergences of these beams differ by several times. The divergence of the SRS in the forward direction increased very slightly when the layered thickness was increased from 8 to 100 mm, but in the backward direction it decreased by a factor of 6 and became smaller than the divergence of the exciting radiation past the focusing lens.

The angular distribution of the SRS in the backward direction is determined to a considerable degree by the structure of the exciting-radiation beam. A narrow beam of the first Stokes SRS component, with sharp edges and with small pulse duration, was formed whenever the exciting laser generated one longitudinal mode and the energy distribution in the beam cross section in the far field was close to Gaussian. The parameters

TABLE 4

Direction	8 mm	20 mm	50 mm	100 mm
"Forward"	132'±7'	143'±8'	177'±2'	180'±4'
"Backward"	168'±8'	93'±5'	31'±2'	29'±2'

48

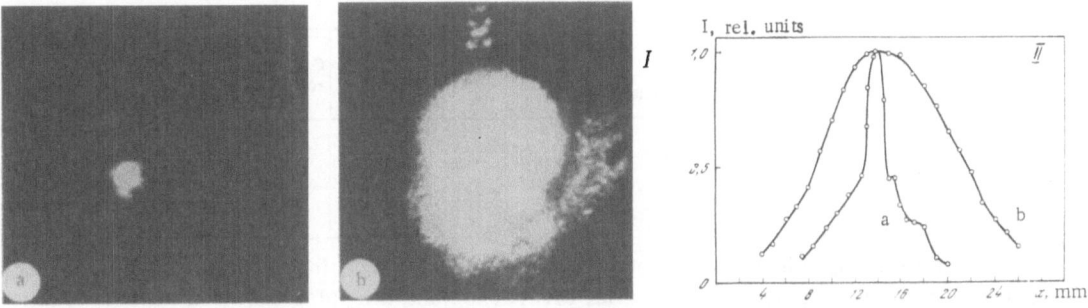

Fig. 7. Energy distribution in the cross section of the beam of the first Stokes SRS component in a liquid-nitrogen layer 100 mm thick in the backward (a) and forward (b) directions. (I) Photograph; (II) results of photometry. The exciting-radiation power density is 2.4 GW/cm^2.

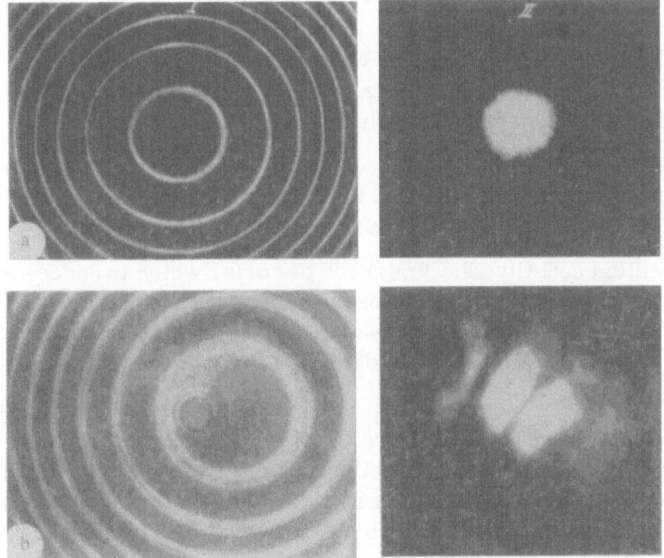

Fig. 8. Energy distribution in the interferogram of laser radiation (I) and in the beam of the first SRS Stokes component in the backward direction in a layer of liquid nitrogen 100 mm thick in the far zone (II) when the SRS is excited by single-mode radiation (a) and by a radiation containing two modes (b).

of the beam of the backward first Stokes component were close in this case to the spatial parameters of the exciting-radiation beam incident on the medium.

It should be noted that in the case when the laser generated two longitudinal modes two maxima were observed in the angular distribution of the energy of the first SRS Stokes component in the backward direction. These maxima were not so pronounced in the angular distribution of the energy of the exciting radiation. Figure 8 shows interferograms of the emission of a ruby laser that generates one or two modes, and also the corresponding distributions of the energy in the far field for the backward SRS.

5. Discussion of Experimental Results and Conclusions

Our experimental results showed that the asymmetry in the formation of the SRS in the direction of the exciting radiation and in the opposite direction is not confined to a difference between the SRS energies in these directions but extends also to other characteristics of the SRS (duration, beam divergence, energy distribution in the beam cross section).

The characteristics of the SRS of the light in the forward and backward directions were closest in liquid nitrogen at a small thickness of the scattering layer near the SRS threshold. This agrees with the theoretical

TABLE 5

Type of radiation	E, J	τ, nsec	P, W	θ, deg	$\Omega \cdot 10^4$, units of solid angle	P/Ω, GW/unit of solid angle	$\Delta\omega$, cm^{-2}
Exciting radiation	0.015	20	$7.5 \cdot 10^6$	0.78	1.4	52	0.015
1St forward	0.018	30	$6 \cdot 10^5$	3.0	21.5	0.28	0.053
1St backward	0.004	9	$4 \cdot 10^5$	0.5	0.6	6.7	0.043

premises [35]. With increasing thickness of the scattering layer and with increasing exciting power, the brightness of the backward SRS increases, and the spatial divergence and the pulse duration decrease. The ratio of the SRS energies in the forward and backward directions changes in this case in a wide range, especially if competing nonlinear effects take place in the medium in addition to the SRS.

By way of illustration of the observed asymmetry of the SRS in liquid nitrogen, Table 5 lists the results of the measurement of the principal characteristics of SRS of light in the forward and backward directions at an exciting power density 0.9 GW/cm^2 and a scattering-layer thickness 100 mm. The table lists the energy E, the pulse duration τ, the power P, the beam divergence in degrees θ, the beam divergence in units of the solid angle Ω, the radiation brightness P/Ω, and the line width $\Delta\omega$. The line width was measured with a Fabry–Perot etalon with a base of 30 mm and with a camera of focal length 840 mm. The error in the measurement of the quantities listed in the table was approximately 10%.

As seen from the table, under these conditions there is formed in the backward direction a beam of the first SRS component close in its spatial structure to the beam of the exciting radiation. The SRS brightness in this direction exceeds the SRS brightness in the forward direction by 25 times.

Thus, a scattering medium constitutes a "mirror" of sorts, which produces simultaneously a frequency shift equal to the most intense molecular vibration of the scattering medium. The restoration of the parameters of the light beam following its SRS, and the compensation for the distortions of its wavefront by the inhomogeneities produced in the medium, take place when definite values of the scattering-layer thickness and of the excitation energy are reached.

A number of papers published in 1965-1967 are devoted to the theoretical and experimental investigation of the SRS energy in the forward and backward directions. Measurements were made of the energy, and sometimes of the SRS pulse duration. The spatial and angular distribution of the energy in the beam of the first SRS Stokes component was not investigated, however. Nor were simultaneous measurements made of all the SRS characteristics. We shall discuss these studies briefly.

Shen and Shaham [45] investigated the connection between the SRS powers in the forward and backward directions with SMBS and with self-focusing of the light. They have observed that the dependences of the SRS power in benzene and toluene on the thickness of the scattering layer and on the excitation energy is different in the forward and backward directions. At a definite cell thickness the radiation power in the backward direction exceeded the power in the forward direction. The asymmetry of the SRS power in toluene increased when a cell with nitrobenzene, a substance with greater self-focusing ability, was placed ahead of the cell with the toluene.

Backward SRS pulses of shorter duration than the forward SRS pulses were observed in a number of studies [46, 47]. Culver, Vanderslice, and Townsend [46] observed shorter pulses of the first SRS Stokes component in hydrogen gas in the backward direction than in the forward direction. Maier, Kaiser, and Giordmaine [47] observed in carbon disulfide pulses of the first Stokes component in the backward direction, of duration $3 \cdot 10^{-11}$ sec, when the SRS was excited with a pulse of duration $12 \cdot 10^{-9}$ sec. The power of the first SRS Stokes component in the backward direction not only exceeded the forward scattering power, but was larger by one order of magnitude than the power of the exciting radiation. The authors of [46, 47] attributed the decrease of SRS pulse duration in the backward direction to its interaction with the opposing pump which has not yet been weakened by transformation into SRS. It was subsequently shown [26], however, that such an interaction can cause only small changes in the pulse duration. The decrease of the pulse duration to several picoseconds is the result of self-focusing of the light in the medium and the formation of moving foci.

In a number of later studies the SRS was excited by picosecond pulses [48, 49]. It was observed that in this case the ratio of the energies of the first Stokes SRS components in the forward and backward directions is not equal to unity [50]. The authors attribute this to the effect of group delay of the pump pulses and of the scattered radiation. This effect, however, should not play any role in excitation of SRS by the nanosecond pulses used in our case.

50

The number of theoretical papers devoted to explanation of the asymmetry of the SRS is quite limited. Bloembergen [32] attributed this effect to the difference between the SRS gains in the forward and backward directions, due to the difference in the damping of the phonons with wave vectors $k_0 + k_S$ and $k_0 - k_S$. However, if this effect were to play a decisive role, then the asymmetry in the liquid nitrogen, where the SRS occurs at 2329.7 cm^{-1}, would be much larger than in carbon disulfide, where the SRS occurs at 656 cm^{-1}. Yet the experimentally observed asymmetry was larger in CS_2 than in liquid nitrogen.

Golger [51] estimated the influence of the SMBS on the energy ratio of the first Stokes SRS component in the forward and backward directions. He considered the formation of the SRS in the presence of advanced SMBS. He obtained expressions for the energy of the first Stokes SRS component in the forward and backward directions, which yielded the ratio

$$\frac{E_f}{E_b} = (1 - a)^{-\frac{\delta P_0}{k P_{thr}} a},$$ (14)

where $a = N_S(0) / N_p(0)$ is the ratio of the photon fluxes of the Stokes component of the SMBS at the exit to that of the pump at the entrance; P_0 is the jump power; $P_{thr} = \delta / g_1$ is the threshold pump; δ is the photon damping; g_1 is the SRS scattering growth rate; g_2 is the SMBS scattering growth rate; $k = (1 - a)g_2 N_p(0)$; the z' axis. is directed from $z = L$ to $z = 0$ (the origin is at the point $z = L$, where L is the thickness of the scattering layer).

At high values of SMBS (for example, in CS_2) this ratio is determined by the excess of the pump over the SRS threshold. At pump values close to threshold, the ratio E_f / E_b is close to unity even at large values of the SMBS, but if the threshold is substantially increased the asymmetry can increase greatly.

If the SMBS exerts a substantial influence on the asymmetry of the SRS energy, then the ratio of the SRS energies in the forward and backward directions should vary with temperature, inasmuch as the SMBS energy decreases when the medium is cooled. However, a role can be assumed here also by other nonlinear effects, so that it is quite difficult to separate the influence of the SMBS. From among all the SRS components investigated by us, the energy ratio in the forward and backward direction increased with decreasing temperature only in the case of the first anti-Stokes component of the SRS in calcite and in carbon disulfide. The possible reason is that the SMBS propagating in the backward direction may play a substantial role in the formation of the first anti-Stokes component in the backward direction. In liquid nitrogen, where we observed no SMBS, the first anti-Stokes component was not observed in the backward direction.

The asymmetry of the SRS parameters in the forward and backward directions is apparently due, on the one hand, to a difference between the propagations of the SRS relative to the exciting radiation (opposing or collinear), and on the other hand to the fact that the SRS is excited not in the entire layer of the scattering medium, and only in a definite thickness of it. Our experiments [36] and the results of other authors [26, 52] have shown that the effective interaction of the exciting radiation with the medium takes place at a definite thickness of the scattering layer near the entrance window of the cell, over the "effective interaction length." (Problems involved in the conversion of the exciting radiation into SRS will be considered in greater detail in Chapter IV.) When short-focus lenses are used to focus the radiation into the medium, the power density of the beam of the exciting radiation changes greatly as it passes through the medium. In this case, the effective interaction length is approximately equal to the distance from the focus to the exit window of the cell. In the case of excitation by a parallel beam, or by a long-focus lens, the effective interaction length is determined apparently by the depletion of the energy of the exciting radiation as it becomes converted into simulated scattering and into other nonlinear effects, and by the saturation of the SRS energy.

In our experiments the exciting radiation was focused into the medium by a lens of focal length 235 mm. The laser-light beam had a diameter of 3 mm on the focusing lens. The radius of the focused spot should amount to $r_F = F\theta$, where F is the focal length of the lens and θ is the divergence of the laser beam.

In our case $r_F = 2.6 \cdot 10^{-2}$ cm. The experimentally determined radius of the circle in the region of the neck (the minimal beam cross section) was $5 \cdot 10^{-2}$ cm. If it is assumed that the caustic is due to spherical aberration of the focusing length, then the length of the region in which the beam is focused is equal to the longitudinal aberration $l = 4\lambda/\alpha^2$, where α is the angular aperture of the beam in the image space. The maximum convergence angle of the radius in our experiments was 10^{-2} rad. Then $l = 2.8$ cm. Thus, in our case there is no clearly pronounced focus, only a narrowing of the beam in the region of the caustic. Therefore the effective interaction length, just as in the case of a parallel beam, is determined by a region in which the exhaustion of the pump and saturation of SRS take place. According to calculations [26, 52], the conversion of the exciting radiation into SRS should occur within a short length of the order of fractions of millimeter. Our investigations of the dependence of the coefficient of conversion into SRS in liquid-nitrogen layers of various thicknesses have shown that at an exciting-radiation density 1 GW/cm^2 the conversion coefficient increases with increasing thickness of the scattering layer to 20 mm, and changes little with further increase of the thickness of the layer. Thus, using a focusing length with F = 235 mm at an exciting-radiation power density 1 GW/cm^2, the effective interaction length was less than 20 mm. If the thickness of the scattering layer is less than the effective interaction length or is equal to it, then the conversion of the exciting radiation in SRS takes place in the entire layer of the medium and the scattering is symmetrical in the forward and backward direction. If the layer thickness exceeds the effective interaction length, then the remaining part of the medium acts as a nonlinear amplifier. As a result, in a definite interval of the power densities of the exciting radiation, the SRS energy in the forward direction can increase with increasing pump more rapidly than in the backward direction. When saturation is reached for the SRS in both directions, the ratio of the SRS energies in these directions becomes close to unity.

In layers whose thickness is less than the effective interaction length, the divergence and distribution of the intensity in the section of the beam of the first Stokes component of the SRS in the forward and backward directions are nearly equal. With increasing thickness of the scattering layer, the divergence of the SRS beam in the forward direction increases, whereas saturation causes the intensity in the central part of the beam to increase more slowly than on the edges.

The formation of the SRS beam in the backward direction is apparently influenced considerably by the interaction of the SRS with the oppositely directed pump. Under certain excitation conditions (when the thickness of the scattering layer exceeds 50 mm and when the power density of the exciting radiation exceeds 1 GW/cm^2), there is formed in the backwards direction a beam of the first SRS Stokes component, close in its spatial parameters (divergence and intensity distribution in the beam cross section) to the parameters of the beam of the exciting radiation incident on the medium. The SRS brightness in the backward direction is then 25 times larger than the radiation brightness in the forward direction, and is close to the brightness of the exciting radiation. It was observed later [53] that the divergence and the spatial distribution of the intensity in the light-beam cross section are restored in the case of SMBS. A detailed explanation of the observed phenomenon was presented by Zel'dovich et al. and interpreted as inversion of the wave front of the light in SMBS. The results of [53] are very similar to the results of our investigations, thus suggesting that the observed phenomena are of the same nature.

To illustrate the reconstruction of the wave front of the laser radiation in SMBS, Zel'dovich et al. [53] introduced into the exciting-light beam a phase plate that distorted its front. Comparing the divergence and the intensity distribution in the SMBS far front passing through the phase plate, and the laser radiation ahead of the plate, they find that the two agree. Without the phase plate, no restoration is observed. In our experiments it is shown that under certain conditions the scattering medium itself restores the wave front.

We have thus observed the restoration of the light-beam parameters in the case of its stimulated Raman scattering in a direction opposite to the direction of propagation of the exciting light, by means of a layer of the medium. The restoration consists in the fact that when definite values of the pump energy and of the layer thickness are reached, the divergence and intensity distribution of the exciting radiation and the SRS differ less in the far field, while the brightness of the backward SRS greatly exceeds the brightness of the SRS in the propagation direction of the exciting radiation.

CHAPTER III

INVESTIGATION OF SELF-FOCUSING OF LIGHT IN SUBSTANCES WITH
DIFFERENT KERR CONSTANTS

1. Survey of the Literature

In the present review we shall dwell on the theoretical and experimental studies of self-focusing of light, published by the time that our research was initiated. Work performed simultaneously with ours will also be considered in the description of the experimental material and in the discussion of the results.

High-power optical radiation can change the dielectric constant of the medium, and consequently also the refractive index. The dependence of the refractive index on the incident-radiation electric-field intensity is expressed in the form

$$n = n_0 + n_2 E^2 + \ldots,$$

(15)

where n_0 is the linear refractive index; n_2 is the nonlinear refractive index. The refractive index has a maximum on the beam axis, where the field intensity is maximum. This causes the rays to deflect towards the beam axis. If the contribution of the nonlinear term n_2E^2 to the refractive index is large, then the refraction of the rays towards the center can offset the diffraction divergence of the beam. This leads to narrowing of the beam and to its self-focusing.

The dependence of the refractive index and the intensity of the incident light wave can be due to various processes that occur in the medium. In liquids with large Kerr constant application of powerful light pulses of approximate duration 10^{-8} sec can cause this dependence to be due to the Kerr effect [22, 24, 54]. It is known that an electric field produces birefringence in various media, which are isotropic in the absence of a field. It can be shown [55] that in a substance whose molecules have no constant moment, the optical field of the laser beam produces the same effects as a static field. The relaxation time for the Kerr effect due to molecular rotation amounts to 10^{-11}-10^{-12} sec.

The nonlinear refractive index connected with the Kerr effect is given by $n_2 = {}^2\!/_3 B\lambda$ (B is the high-frequency Kerr constant) [24]. The high-frequency Kerr constant can be measured directly by registering the birefringence produced under the influence of the light-wave field [55-57], or can be calculated if the Kerr constant in the constant field is known [58]. The high-frequency Kerr constant is equal to

$$ B = \frac{(n_0^2+2)(n_0^2-1)(\varepsilon+2)}{(\delta+2)^2(\delta-1)}\,K, \tag{16} $$

where δ is the dielectric constant and ε is given by the Debye relation

$$ \frac{\delta-1}{\delta+2} = \frac{\varepsilon-1}{\varepsilon+2} + \frac{4\pi\rho_0\mu^2}{9mKT}, \tag{17} $$

K is the Kerr constant in a constant field, and μ is the constant dipole moment of the molecule. For nonpolar molecules $K \approx B$, whereas for polar molecules with large μ the high-frequency Kerr-effect constant can be much less in absolute magnitude than in a constant field.

The self-focusing of light in a medium can also be due to the nonlinearity due to electrostriction. The nonlinear refractive index n_2 for the case of electrostriction is determined by the expression [16]

$$ n_2 = (\gamma^2/16\pi)\,n_0B', \tag{18} $$

where $\gamma = \rho\,d\varepsilon/d\rho$, ρ is the density; $B' = \rho\,dp/d\rho$ is the volume compression modulus; p is the pressure.

The values of the threshold power and of the self-focusing length (the distance over which the beam collapses into a thin filament) can be determined by solving the electromagnetic wave equation that takes into account the nonlinearity of the medium [24, 54]:

$$ \nabla^2\mathbf{E}^2 - \frac{\varepsilon_0}{c^2}\frac{\partial^2\mathbf{E}}{\partial t^2} - \frac{\varepsilon_2}{c^2}\frac{\partial^2}{\partial t^2}(E^2\mathbf{E}) = 0, \tag{19} $$

where ε_0 and ε_2 are real quantities and $\varepsilon_2E^2 \ll 1$.

Assume that a linearly polarized wave of frequency ω propagates along the z axis:

$$ \mathbf{E} = \frac{\hat{\rho}}{2}\,(E'e^{i(kz-\omega t)} + \text{c.c.}), \tag{20} $$

where $k = \varepsilon_0^{1/2}\omega/c$, the exponential $\exp(ikz - i\omega t)$ characterizes the wave propagation, and E' is the slowly varying amplitude. Substituting this expression into the wave equation, neglecting terms containing the third harmonic, and omitting a small term that contains the second derivative of E' with respect to z, we obtain

$$ 2ik\frac{\partial E'}{\partial z} + (\nabla_x^2 + \nabla_y^2)E' + \frac{\varepsilon_2'k^2}{\varepsilon_0}|E'|^2 = 0. \tag{21} $$

If there are no transverse gradients in the beam, then this equation has a simple solution:

$$ E' = E_0'e^{in_2'k|E_0'|z/n_0}. \tag{22} $$

This solution can be used to determine the self-focusing length. If the transverse second derivatives of E' depend on the distance along the z axis (via the transverse change of the intensity in the argument of the exponential in the expression for E'), then the field changes noticeably along the axis if

$$ z \sim z_{\text{foc}} \equiv \frac{a}{2P_m'}\left(\frac{n_0}{n_2'}\right)^{1/2}, \tag{23} $$

where a is the characteristic transverse radius of curvature of the entering beam (a coincides with the actual radius of the beam if the distribution of the intensity in the cross section has a parabolic profile); P'_m is the peak power; $n'_2 = n_2/2$. After introducing corrections for diffraction, we obtain the self-focusing length

$$z_{foc} = \frac{a}{2}\,(n_0/n'_2)^{1/2}\,(P'_m - P_{cr})^{-1}, \tag{24}$$

where $P_{cr} = 1.22\lambda/8a(n'_2 n_0)^{1/2}$. Equating z_{foc} to the diffraction length, we obtain an approximate value of the threshold power [54, 59]

$$P_{cr} = \frac{c\lambda^2}{8\pi^2 n_0^2 n_2}\,, \tag{25}$$

where λ is the wavelength.

Similar expressions for the critical power and the self-focusing length were obtained also by Lugovoi [26, 60].

The analytic results do not contain a number of significant singularities of the interaction of the light beam with the nonlinear medium. The picture of the propagation of the beam passed the "collapse" point was not considered. It was assumed that past this point self-capture of the beam into the waveguide regime takes place. A detailed theoretical analysis of the kinetics of self-focusing was carried out by Prokhorov and Lugovoi and their co-workers. The equations describing the propagation of high-power laser radiation in a nonlinear medium were solved numerically. They have shown that the self-focused laser beam breaks up into annular zones which are focused into points located on the beam axis [25, 26]. At an exciting-radiation power close to threshold, one point is formed. At high power, an number of foci are produced in the medium and can move both in the propagation direction of the exciting radiation and in the opposite direction, depending on the shape of the radiation pulse. The dimensions of the focal regions and the distances between them depend on the form of the nonlinear absorption in the medium. The transverse dimensions of the focal regions increase with increasing number of the focus, and remain at the same time similar to each other. In self-focusing of picosecond pulses, the points can become divided and move in opposite directions.

These theoretical concepts were confirmed in a large number of studies. Korobkin and co-workers [27] obtained photographs of the trajectories of the foci in a multifocus structure using a high-speed time scan of the investigated processes. Loy and Shen [61] have observed that the experimental values of the distance between the moving focus and the ultrashort backward SRS pulse coincide for various instants of time with the theoretical ones calculated from the theory of moving foci. Bright points produced at the location of the turning of the foci of a multifocus structure were observed in [27] for the case of picosecond pulses propagating in the medium.

In the first experiments and theoretical studies devoted to self-focusing of laser radiation, the SRS was either disregarded or considered as an effect that leads to loss and disintegration of the light-conducting filaments [62]. On the other hand, in the investigations of the regularities of the formation and development of the SRS it was assumed that SRS in substances with large Kerr constants arise and propagate in the self-focusing filaments of the exciting radiation [63, 64]. This was taken to be the cause of the deviations of the experimental results from theory, which were observed in investigations of SRS.

Practically all the studies dealing with the connection between SRS and self-focusing were aimed at proving this concept. As a result, attention was focused on substances with clearly pronounced self-focusing ability (carbon disulfide, nitrobenzene). Substances with small Kerr constants were not investigated systematically. It was assumed that there is no self-focusing in these substances and of all the rules governing the development of the SRS agree with the theory.

The experimental study of self-focusing has been the subject of a large number of investigations. In the present review we dwell only on those dealing with the connection between self-focusing and the SRS phenomenon. In 1965-1967, a number of studies were made [13-15, 65] in which the SRS gain was measured and compared with the theoretical values. It turned out that good agreement is obtained only in hydrogen gas [65]. In liquids, on the other hand, the gain turned out to be larger by one or two orders of magnitude than predicted by the theory [13, 32]. Many workers explained this anomalous behavior of the SRS gain as being due to self-focusing of the radiation in the medium and to formation of light-carrying filaments [28, 32, 55]. The appreciable increase in the density of the exciting-radiation power in the filaments could lead to a lowering of the

SRS threshold and to an increase of the gain. This assumption was experimentally confirmed in a number of studies. In particular, it was shown [32] that the SRS gain depends on the length of the cell. The gain increased rapidly at a certain definite length, and this length increased with decreasing exciting radiation. Bret and Denariez [18] investigated the dependence of the energy of the first Stokes SRS component in acetone on the energy of the exciting radiation. The experiments were performed both on pure acetone and on a mixture containing 90% acetone and 10% carbon disulfide. It turned out that in the mixture with the carbon disulfide the plot of the energy of the first Stokes SRS component of the acetone on the excitation energy experiences a jump of three orders of magnitude, whereas no jump is observed in pure acetone. The authors believe that the cause of the jump is the increase of the gain as a result of the self-focusing of the exciting radiation produced in the CS_2.

At the same time, investigations were made of the influence of self-focusing of the exciting light on the SRS threshold. It should be noted that the SRS threshold is to a certain degree an arbitrary quantity. The concept of threshold is applicable only under rigorously fixed experimental conditions, including the conditions for the registration of the SRS. However, relative measurements of the SRS threshold can yield some idea of how the SRS energy changes with changing excitation conditions. In an investigation of the influence of self-focusing on the SRS threshold, cells of different length, with different focusing substances, were placed ahead of the cell with the investigated substance. The SRS threshold was measured in the second cell. Shen and Shaham [28], who used this method, observed that at a certain length of the first cell the SRS threshold of the investigated substance in the second cell decreased abruptly. They assumed that this is due to the onset of self-focusing of the exciting radiation in the first cell. Lallemand and Bloembergen [14] have described a number of experiments that show that the SRS threshold depends strongly on the focusing properties of the medium. They have observed that if a cell with bromobenzene is placed ahead of a cell with nitrobenzene, then the threshold length for the onset of SRS in nitrobenzene decreases by a factor of 10. The decrease of the threshold length took place even before the SRS appeared in the first cell. It was also observed that it is possible to excite SRS in liquids in which the threshold is not reached, if these liquids are dissolved in strongly focusing liquids. Wang [66], measuring the dependence of the SRS threshold on the length of the cell for various liquids and extrapolating these data to a cell of infinite length, obtained the values of the critical power of self-trapping of the beam in a liquid. Maier and Kaiser [29] have shown experimentally that the SRS threshold power is inversely proportional to the Kerr constant. This should be expected from the expression (25) for P_{cr} if the appearance of the self-focusing of the exciting light leads to appearance of SRS.

Thus, by the time the present study was started, questions connected with the mutual influence of SRS and self-focusing of the exciting light in substances had attracted great interest. However, the investigations up to that time did not yield a complete picture of the connection between these two nonlinear effects in a medium.

Self-focusing of the exciting radiation was investigated as a rule by indirect methods, by determining the change of the SRS threshold. Even when the picture of self-focusing light was photographed directly inside the medium [32], no spectral separation of the exciting radiation from the SRS was made. Substances with small Kerr constants were hardly investigated.

The aim of the present study was to investigate in detail the development of SRS and self-focusing of light in substances with different self-focusing abilities. It was proposed to investigate the influence exerted on the self-focusing regime by the power density of the exciting radiation, by the thickness of the scattering layer, and by the temperature of the scattering medium.

2. Procedure of Investigation of the Self-Focusing of Light

Self-focusing of light is accompanied by a change in the structure of the light beam. Observations inside the medium reveal in the cross section of the light beam distinct small-diameter spots that do not appear in the light beam incident on the cell. In the present study, self-focusing spots were observed in planes near the entrance and exit windows of the cell or near the entrance and exit faces of the crystal inside the medium. Spectral separation of the components was carried out with the aid of a stack of selective filters or with a spectrograph. The energy distribution in the cross sections of the exciting-radiation beam or of the SRS beam were photographed with 16 × magnification. This made it possible to determine the self-focusing threshold, the number and diameter of the self-focusing spots, their distribution in the field of radiation of the undisplaced component and of the SRS components. These investigations were made simultaneously with temporal, spatial, and energy measurements of the SRS, so that we were able to establish the connection between different characteristics of SRS with the self-focusing of light.

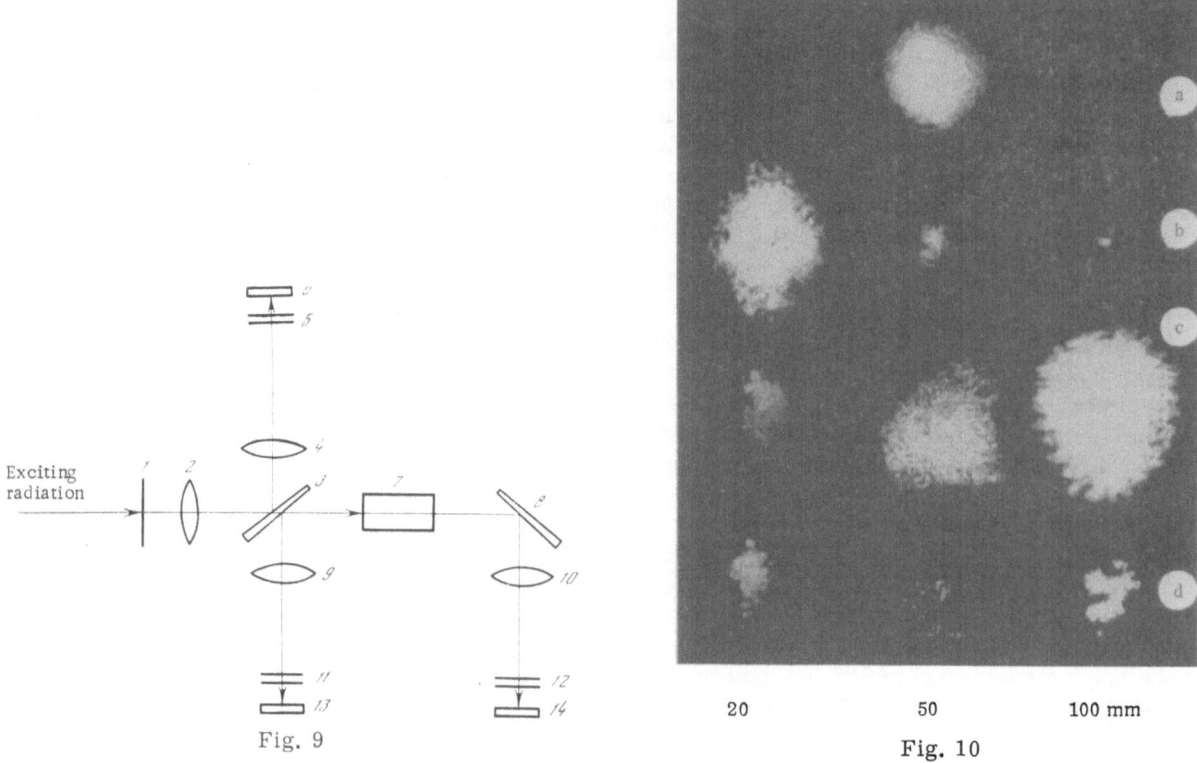

Fig. 9

20   50   100 mm

Fig. 10

Fig. 9. Optical diagram of the setup for the investigation of self-focusing of light. (1, 5, 11, 12) Light filters; (2, 4, 9, 10) lenses; (3, 8) turning wedges; (7) cell with medium; (6, 13, 14) photographic plates.

Fig. 10. Cross sections of exciting-radiation beam at the entrance to the medium (a) and at the exit from a dewar with liquid nitrogen (b), and of the beam of the first SRS Stokes component inside the medium near the exit (c) and entrance (d) windows of the dewar at various thicknesses of the scattering layer. The exciting-radiation power density was 1 GW/cm^2.

The experimental setup is illustrated in Fig. 9. The exciting radiation, whose energy could be varied with neutral light filters 1, was focused by lens 2 of focal length 235 mm into the investigated medium 7. The lenses 4, 9, and 10 projected the images of the cross section of the exciting-radiation beam entering the medium and of the planes inside the medium near the entrance window of the cell (radiation in the backward direction) and near the exit window of the cell (radiation in the forward direction), with 16 × magnification, on photographic plates 6, 13, and 14. The beams of the scattered and exciting light were reflected by wedge-shaped glass plates 3 and 8, to prevent the doubling of the image. The necessary components were separated with the aid of glass or interference light filters 5, 11, and 12, which were placed in front of the photographic plates. In a number of cases the picture of the self-focusing of the exciting light and of the SRS components was registered with a spectrograph. To this end, the planes inside the cell near the front and rear windows were projected by lenses onto the plane of the spectrograph slit. The slit was removed in this case. The photograph showed the distribution of the intensities of various SRS components inside the medium in the beam cross section. The dimensions of the self-focusing spot were determined by photometry of the image of the spot, the photographic density was converted into intensity with the aid of the density curve, and after constructing the contour of the image we measured its diameter. Taking into account the magnification of the optical system, we were able to obtain the true diameter of the filament. The resolving power of this system was determined with the aid of a set of targets. The target was placed at the location of the exit window of the cell, illuminated with a gas laser, and its image was photographed. The resolving power of the optical system, determined from the distance between the lines of the most sharply focused image, was 10 μm. By displacing the target and photographing its image we found the depth of focus to be 1 mm.

3. SRS Self-Focusing of Light in Liquid Nitrogen

An analysis of previous studies of self-focusing of light shows that they were aimed primarily on determining the structure of the laser beam after passing through the medium. In the case when a structure of the

SRS Stokes-component beams was observed simultaneously with the exciting radiation, the authors noted that each self-focusing point in the section of the exciting-radiation beam corresponds to a self-focusing point in the section of the SRS beam. This served as proof of the appearance and propagation of SRS in the self-focusing filaments of the exciting radiation. Investigations of this type were carried out in carbon disulfide and nitrobenzene.

In our studies we chose liquid nitrogen as the object of the investigation. It is known that liquid nitrogen has a low threshold for SRS of light (lower than carbon disulfide). The Kerr constant of liquid nitrogen is low, and it was assumed therefore for a long time that no self-focusing takes place in this substance [33]. In the investigation of the intensity distribution in the sections of the exciting-radiation beam and of the first-Stokes SRS component beam in liquid nitrogen we observed, for the first time ever, a characteristic self-focusing pattern in the beam of the first Stokes SRS component in the absence of self-focusing of the exciting radiation [36, 67-69]. The threshold of self-focusing of SRS in liquid nitrogen practically coincided with the onset of the SRS. At a scattering-layer thickness 50 mm, the minimum power density of the exciting radiation, at which we were able to register and observe self-focusing of SRS, was 80 MW/cm^2. The self-focusing of the exciting radiation in such layers started only at a power density 1 GW/cm^2. The pictures of the self-focusing of the exciting radiation and of SRS were different. Figure 10 shows a magnified picture of the intensity distribution in the cross section of the exciting-radiation beam near the exit window of an empty dewar (Fig. 10a) and of a dewar filled with liquid nitrogen (Fig. 10b), as well as the cross section of the beam of first SRS Stokes component in liquid nitrogen near the exit (Fig. 10c) and entrance (Fig. 10d) windows of the dewar inside the medium at various thicknesses of the scattering layer. The power density of the exciting radiation was 1 GW/cm^2. It is seen from the figure that the distribution of the energy in the beam cross section is different for the exciting radiation and for the SRS. At a layer of thickness 50 mm, the exciting-radiation beam decreased in size from 1 mm to 300 μm, whereas for the first Stokes component spots of 10 μm diameter (the resolution limit of our optical system) were observed. At a scattering-layer thickness 100 mm, the beam of the exciting radiation decreased to 50 μm. In the beam of the first Stokes component, on the contrary, the spots increased somewhat with increasing thickness of the scattering layer from 50 to 100 mm. Whereas in the beam of the exciting radiation we observed one point, in the beam of the first Stokes SRS component there were several.

The intensity distribution in the cross section of the beam of the first Stokes SRS component inside the medium near the entrance and exit windows of the dewar was approximately the same at small thickness of the scattering layer. As seen from Fig. 10, at an excitation-power density 1 GW/cm^2, the picture was approximately the same in the forward and backward directions in a layer 20 mm thick. With increasing layer thickness, the picture became asymmetrical: whereas at the entrance window (in the backward direction), at all investigated scattering-layer thickness, self-scattering points were observed, at the exit face (in the forward direction) the distribution of the intensity in the beam was diffuse when the layer thickness exceeded 50 mm.

At low power density of the exciting radiation, the structures of the beams of the first SRS Stokes component propagating in the excitation direction and in the opposite direction were the same. With increasing excitation power density, the distribution of the energy in the cross section of the beams in the forward and backward directions became different. Figure 11 shows the cross section of the beam of the first Stokes SRS component in a liquid nitrogen layer 50 mm thick near the entrance (backward) and exit (forward) windows at different power densities of the exciting radiation. When the power density of the exciting radiation was 80 MW/cm^2, one point with dimension 30-50 μm was observed in the SRS, sometimes against the background of a weaker spot with dimension 150 μm. The picture was the same both near the entrance window and near the exit window of the dewar. With increasing exciting-radiation power density, the number of spots increased. In the region from 150 to 300 MW/cm^2, a jump is observed in the number of points, and a decrease of their diameter to 20-25 μm. Simultaneously with this jump, a sharp increase took place in the intensity of the first Stokes SRS component. At an exciting radiation power density exceeding 350 MW/cm^2 the picture became asymmetrical: a large number of minute dots (with dimension 10 μm) were observed at the entrance face, while at the exit the distribution of the intensity in the field of the weak Stokes SRS component was uniform.

The intensity distribution in the exciting-radiation beam was uniform at low power density of the exciting radiation, and narrowing of the beam set in only at 1 GW/cm^2 (see Fig. 10b).

The investigations have revealed SRS self-focusing in a substance with small Kerr constant. Its threshold is lower than the self-focusing threshold of the exciting radiation, and it takes place independently of the self-focusing of the exciting radiation. The experimental results indicate that the refractive-index change that leads to the self-focusing of the SRS of light in liquid nitrogen can be due not to the quadratic Kerr effect, but to some other process. Our further investigations were aimed at a detailed study of the phenomenon of SRS self-focusing.

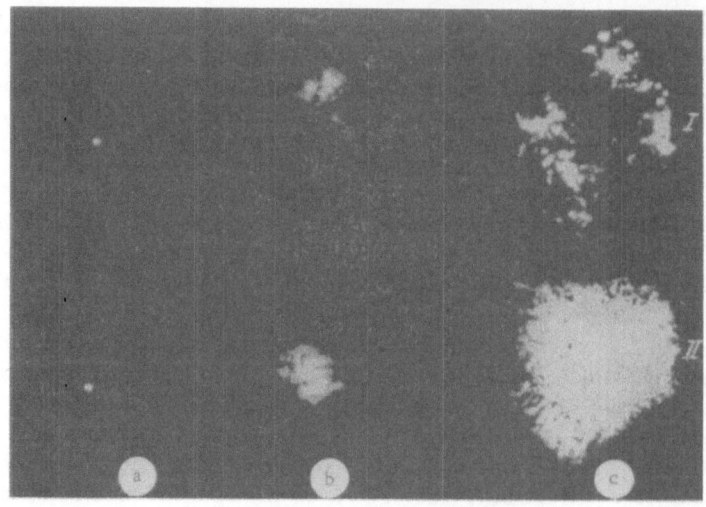

Fig. 11. Cross section of beam of first SRS Stokes component in a layer of liquid nitrogen 50 mm thick, in the interior of the substance, near the entrance (I) and exit (II) windows of the dewar at various exciting-radiation power densities. (a) 80 MW/cm^2; (b) 250 MW/cm^2; (c) 1200 MW/cm^2.

4. Self-Focusing of SRS in Calcite

We observed also self-focusing of SRS in single-crystal calcite. The authors of the first studies of SRS believe that the SRS phenomenon in single-crystal calcite is not encumbered by such nonlinear effects as self-focusing, so that this phenomenon should satisfy the theoretical relations. Self-focusing of the exciting radiation was in fact not observed in the entire investigated power-density interval (up to 1 GW/cm^2) and temperature interval (from −196 to +150°C), but only starting with an exciting-radiation power density 350 MW/cm^2. At a lower power density of the exciting radiation, the sensitivity of our apparatus was insufficient to register the SRS on a photographic plate.

Figure 12 shows the cross section of the beam of the first SRS Stokes component in a calcite single crystal of thickness 10 mm near the entrance and exit faces of the crystal, at various exciting-radiation power densities. With increasing excitation power density, the number of points increased, and their diameter decreased. The self-focusing picture was approximately the same in the forward and backward directions, up to an excitation power density of 1 GW/cm^2. At higher power densities, the number of points at the exit face became larger than at the entrance face. At an excitation power density higher than 1.2 GW/cm^2, the intensity distribution in the SRS field near the exit face became diffuse.

A change in the temperature of the calcite single crystal also changed the picture of the self-focusing of the SRS. Figure 13 shows the intensity distribution in the field of the first SRS Stokes component in calcite near the entrance face of the crystal at temperatures −196 and +20°C. As seen from the figure, lowering the temperature with the remaining excitation conditions constant led to an increase in the number of SRS self-focusing points and to a decrease in their diameter.

5. Self-Focusing of Light in Carbon Disulfide and Nitrobenzene

Carbon disulfide and nitrobenzene are substances with large Kerr constants. Self-focusing in these substances, especially in carbon disulfide, has been the subject of many studies. It was assumed that SRS in such substances as carbon disulfide and nitrobenzene is produced and propagates in laser-radiation self-focusing filaments and that its threshold coincides with the self-focusing threshold of the exciting light [70-72]. In this case the position of the self-focusing dots in the SRS field should be the same as in the field of the exciting radiation. In the present study we observed in carbon disulfide and nitrobenzene self-focusing of both the exciting radiation and of the SRS [38, 73]. The minimum dimension of the self-focusing dots was 10 μm (the resolution limit of the optical system). However, despite the previously prevailing notions, the picture of the self-focusing of SRS and of the self-focusing of the exciting radiation was not always the same. Figure 14 shows the cross section of the beam of the undisplaced component and of the first Stokes component of SRS at the entrance window of a cell with carbon disulfide 50 mm long at various densities of the exciting radiation. No

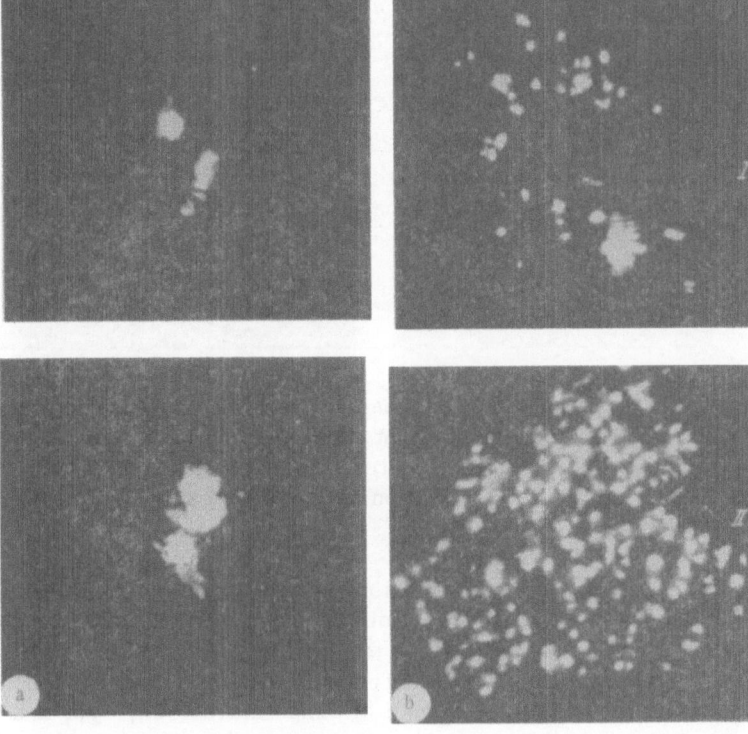

Fig. 12. Cross section of SRS beam in a calcite single crystal 10 mm thick, inside the medium near the entrance (I) and exit (II) faces of the crystal at various exciting-radiation power densities. (a) 500 MW/cm^2; (b) 1200 MW/cm^2.

self-focusing took place at a power density higher than 1.5 GW/cm^2, whereas for the first Stokes component of the SRS self-focusing was observed. With decreasing exciting-radiation power density, self-focusing appeared not only in the first SRS Stokes component but also in the radiation at the undisplaced frequency. A similar phenomenon was observed by Loy and Shen [72] at the exit window of the cell with CS_2, and by Shen and Shaham in benzene [45].

For a cell 20 mm long, the picture of self-focusing in CS_2 is the same for the Stokes component and for the undisplaced component at an excitation power density less than 1 GW/cm^2. Thus, our experiments have shown that, depending on the excitation power density and on the thickness of the scattering layer, the picture of the distribution of the intensity in the SRS field in CS_2 can either duplicate the distribution of the intensity in the exciting-radiation field, or can substantially differ from it.

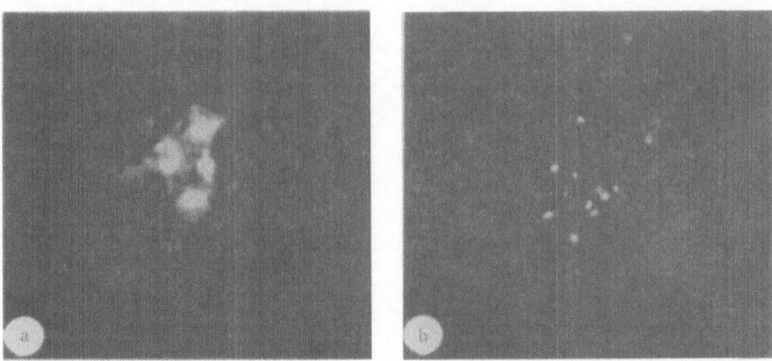

Fig. 13. Cross section of beam of first Stokes SRS component inside a calcite single crystal near the exit face at temperature +20°C (a) and −196°C (b) and at a power density 100 MW/cm^2.

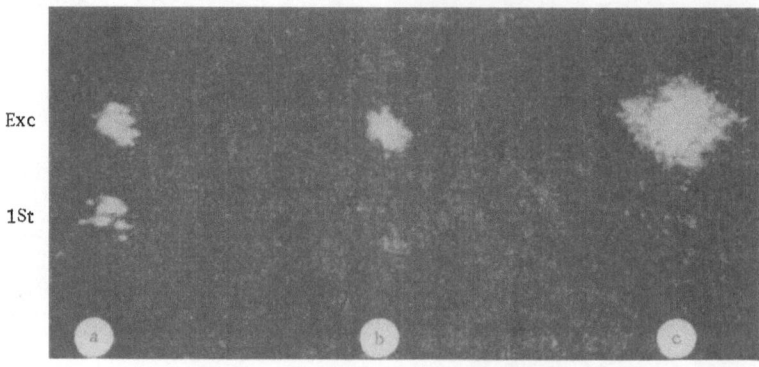

Fig. 14. Cross section of the beam of the exciting radiation (Exc) and of the first Stokes component (1St) of SRS in carbon disulfide within the substance near the entrance window of a cell 50 mm long at various exciting-radiation energies. (a) 38 MW/cm^2; (b) 100 MW/cm^2; (c) 1500 MW/cm^2.

The change of the temperature of the medium also influenced the picture of the self-focusing in CS$_2$. The change of the structure of the beams of the exciting radiation and of the SRS with changing temperature in CS$_2$ is shown in Fig. 15. The photographs were taken at an exciting-radiation power density 1 GW/cm^2 with a cell of 20 mm length. The figure shows the field of the exciting radiation and of the first and second Stokes components inside the medium near the entrance window of the cell at temperatures $+20$, -70, and $-90°$C. At all temperatures the field picture is the same for all components. At room temperature, the distribution of the intensity in the field takes the form of either a diffuse spot or of a ring, against the background of which a number of dots is sometimes observed. At a temperature from -60 to $-70°$C one can see in the beam cross section self-focusing dots of 10 μm diameter, arranged along a ring or over the entire beam cross section. With further lowering of the temperature, the dots coalesce into one spot of somewhat larger diameter than the filament.

In nitrobenzene we have likewise observed self-focusing of both the SRS and of the exciting radiation. Whereas near the entrance window of the cell the self-focusing picture was approximately the same for the SRS and the exciting radiation, at the exit from the cell the fields of the SRS and of the exciting radiation differed greatly. The SRS field consisted of distinct sharp dots, whereas for the exciting radiation weak dots were observed against a diffuse background. The self-focusing of the exciting radiation is shown in Fig. 16. Figure 17 shows the field of the first SRS Stokes component in nitrobenzene near the entrance and exit windows of the

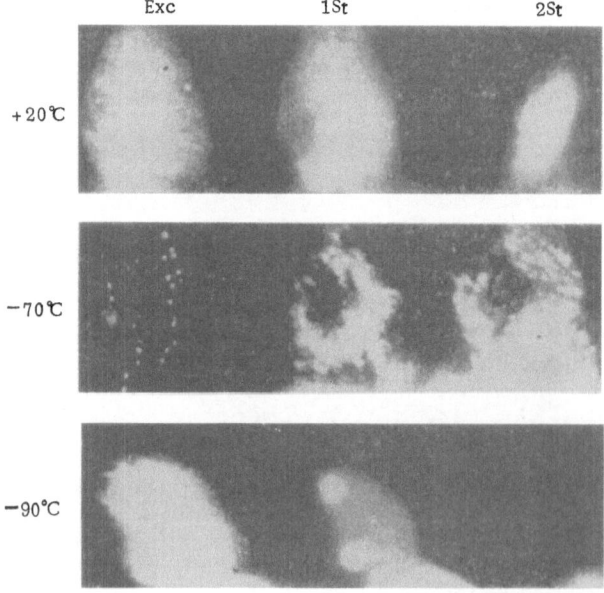

Fig. 15. A cross section of exciting-radiation beam (Exc) and of the first and second Stokes components of SRS in carbon disulfide inside the medium near the entrance window of the cell, at various temperatures. The image is doubled because of reflections from the two faces of the turning plate; the density of the exciting radiation power is 1 GW/cm^2.

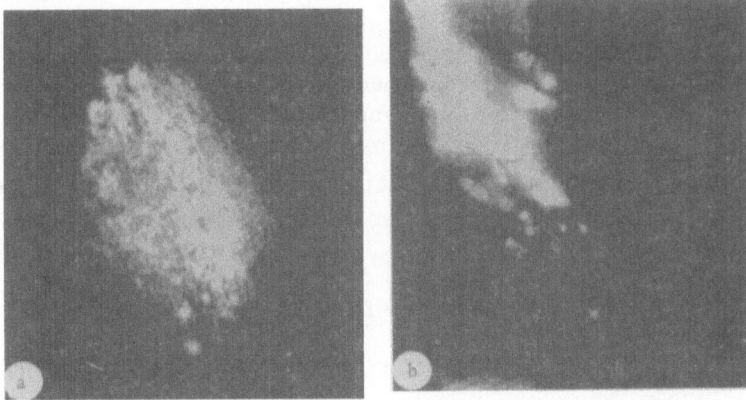

Fig. 16. Distribution of the intensity in the field of the exciting radiation in nitrobenzene in the interior of the medium near the exit (a) and entrance (b) windows of the cell. Layer thickness 50 mm, temperature +20°C. Exciting-radiation power density 1 GW/cm².

cell at different exciting-radiation power densities. In the forward direction (near the exit window) we could register SRS in benzene starting with an exciting-radiation power density 60 MW/cm². A dot of 65 μm diameter was observed in this case in the SRS field. Only at an excitation power density 700 MW/cm² did the number of dots increase and their diameter decrease. Near the entrance window of the cell we were able to observe SRS only at power densities higher than 300 MW/cm².

In the SRS field near the entrance window we always observed several dots. Their number increased and their diameter decreased with increasing excitation power. The self-focusing dots near the entrance window were more loosely arranged, over a larger surface, whereas near the exit window they formed a more compact spot. Both in the exciting-radiation field and in the field of the first Stokes component of the SRS, one or several systems of concentric rings were superimposed on the picture of the dots.

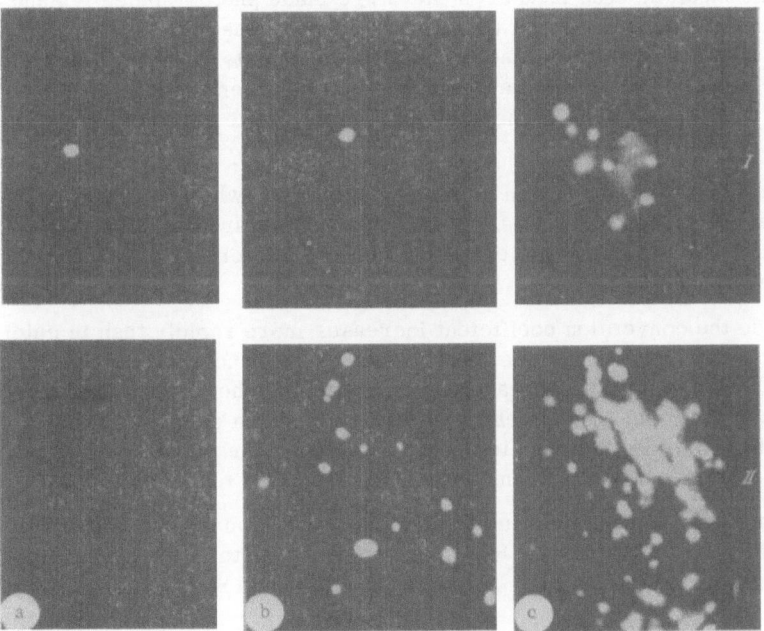

Fig. 17. Cross section of beam of first Stokes SRS component in nitrobenzene in the interior of the medium near the exit (forward — I) and entrance (backward — II) window of the cell at various exciting-radiation power densities. (a) 60 MW/cm²; (b) 300 MW/cm²; (c) 1200 MW/cm².

The difference between the pictures of the self-focusing of the exciting radiation and that of the first Stokes component of the SRS under definite excitation conditions indicates that the development of the SRS and of the self-focusing of the SRS of light in these substances can proceed independently of the self-focusing of the exciting radiation. In contrast to substances with small Kerr constants, in substances with large Kerr constants the significant contribution to the self-focusing mechanism can be made by the Kerr effect.

In all the investigated substances, the number and diameter of the SRS self-focusing dots depends on the power density of the exciting radiation, on the thickness of the scattering layer, and on the temperature of the substance.

CHAPTER IV

EXPERIMENTAL INVESTIGATION OF THE DEPENDENCE OF THE SRS
SELF-FOCUSING ON THE SRS POWER

1. Dependence of the Coefficient of Conversion of the Exciting Radiation into SRS on the Temperature of the Medium

The experimental results of the investigation of the self-focusing of SRS of light, cited in the preceding chapter, have shown that the observed picture of the self-focusing depends on the conditions of the SRS excitation. A change in the excitation conditions, i.e., a change in the exciting-radiation power, in the thickness of the scattering layer, or in the temperature of the medium influences primarily the coefficient of conversion of the exciting radiation into SRS of light. It was of interest in this connection to investigate the dependence of the regime of the SRS self-focusing on the power converted into the first SRS Stokes component. An investigation of the power of the first SRS Stokes component and of the coefficient of conversion into SRS is an independent problem, to which a large number of papers is devoted [14, 18, 33, 45, 74-79]. We therefore do not claim here a complete discussion of this question, and regard the investigations of the conversion coefficients in our paper as auxiliary.

We investigated the dependence of the coefficient of conversion of light into SRS on the temperature in calcite and in carbon disulfide [39, 40]. We measured the ratio of the SRS energy to the ratio of the exciting radiation incident on the medium. The measurements were made photographically, and the necessary SRS components were separated with a spectrograph. The energies of the SRS components were obtained as a result of averaging over the data of 15-20 measurements. Table 6 lists the coefficient of conversion of the exciting radiation into SRS in carbon disulfide and calcite at various temperatures. The exciting-radiation power density is 2 GW/cm^2, and the thickness of the scattering layer of the carbon disulfide is 20 mm, and that of calcite is 10 mm.

As seen from the table, the conversion coefficient increases when the medium is cooled, both in carbon disulfide and in calcite, and then this increase slows down and in carbon disulfide at $-90°C$ one observes even a certain decrease of the conversion coefficient into SRS. The conversion coefficient is maximal in carbon disulfide at a temperature from -40 to $-60°C$.

In carbon disulfide the conversion coefficient increases more rapidly than in calcite. In the temperature interval where it reaches the maximum value, the intensity of the exciting radiation passing through the medium decreases. In carbon disulfide it decreases by a factor of 6 when the temperature of the medium is decreased from -40 to $-70°C$. In calcite, the change of the conversion coefficient and of the exciting-radiation intensity with temperature is smoother than in carbon disulfide. The intensity of the exciting radiation decreases to approximately one-half when the medium is cooled from $+100$ to $-100°C$.

We investigated not only the combined coefficient of conversion of the exciting radiation into SRS, but also the distribution of the energy over the SRS components in calcite and in carbon disulfide at various temperatures. In Table 7 are listed the energies of the first and second Stokes components (1St, 2St) and of the

TABLE 6

t, °C	Carbon disulfide				Calcite				
	+20	−40	−75	−90	+150	+20	−20	−110	−196
Conversion coefficient, %	8	59	54	46	23	33	56	76	76

TABLE 7

Component	t, °C				
	+20	—30	—40	—75	—90
1St	24 000	170 000	154 000	121 000	104 000
1aSt	19.6	52.2	58.7	97.8	52.2
2St	71.8	2220	3530	28 500	26 200
Exciting light	65 200	66 700	80 500	11 900	16 200

TABLE 8

Component	t, °C				
	150	20	—20	—110	—196
1St	57 000	68 000	126 000	197 000	177 000
2St	3	8	17	16	590
1aSt	10	8	8	12	310
2aSt	—	0.01	0.01	0.03	0.85

first anti-Stokes component (1aSt) of the SRS and of the exciting radiation passing through carbon disulfide at various temperatures, at an exciting-radiation density 2 GW/cm^2 and at a scattering-layer thickness 20 mm.

As seen from the table, the greater part of the SRS energy is concentrated in the first Stokes component. With decreasing temperature, an appreciable growth of the energy of all the SRS components is observed, as well as a redistribution of the energy in favor of components of ever higher order. A particularly strong increase occurs in the intensity of the second SRS Stokes components, which changes by more than two orders of magnitude in the temperature interval from +20 to −90°C. For the first Stokes component and for the first anti-Stokes component, at a temperature −70°C, saturation sets in, and at the same time a third Stokes and a second anti-Stokes SRS component appear in the SRS spectrum. At temperatures from −80 to −90°C, some saturation of the combined SRS intensity sets in.

A similar temperature dependence of the intensity of the SRS components was observed in calcite. Table 8 lists the energies of the first and second Stokes and of the first and second anti-Stokes SRS components in calcite. The energy density of the exciting radiation is 2 GW/cm^2 and the thickness of the calcite single crystal is 1 cm.

It is seen from the table that in calcite, just as in carbon disulfide, the energy of the SRS at an exciting-radiation power density 1-2 GW/cm^2 is concentrated mainly in the first Stokes component of the SRS. With decreasing temperature of the medium, the energy of the SRS components increases and with more energy going to the higher-order components.

The variation of the SRS intensity with temperature, as shown in our paper [79], is determined to a considerable degree by the properties of the ordinary Raman scattering. The growth of the cross section of the ordinary Raman scattering with decreasing temperature can lead to an increase of the SRS gain and to an increase of its intensity. Investigations of Raman scattering in carbon disulfide have shown that in the temperature interval from +20 to −100°C the integrated intensity of the 656 cm^{-1} line increases by three times [42], whereas its width decreases from 0.9 to 0.7 cm^{-1} [80]. The integrated intensity of the 1086 cm^{-1} line of calcite, according to the data of Tulub and Bobovich [81], decreases by 30% when the crystal is heated from 20 to 277°C. The width of this line, measured by Park [82], is 1.1 cm^{-1} at room temperature and approximately 0.5 cm^{-1} at 40°K. When the crystal is heated to 670°C, the linewidth increases to 15 cm^{-1} [83]. An important role in the SRS excitation is played by the intensity of the corresponding Raman-scattering line at the maximum, i.e., by the integrated intensity divided by the width. Plots of the temperature dependence of the intensity at the maximum of the lines 656 cm^{-1} of carbon disulfide and 1086 cm^{-1} of calcite, obtained from the cited references, are shown in Fig. 18. The same figure shows plots of the relative intensities of the first Stokes components of the SRS in carbon disulfide and in calcite.

A comparison of the plots shown in Fig. 18 shows a definite correspondence between them: the Raman scattering of carbon disulfide, just as the SRS, varies much more strongly with temperature than the Raman scattering and the SRS of calcite. Since the second Stokes components are the results of successive excitation or of parametric interaction of lower-order components, the temperature dependence of the intensity of the second Stokes components is determined to a considerable degree by the temperature dependence of the first

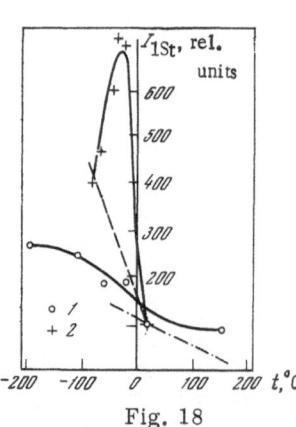
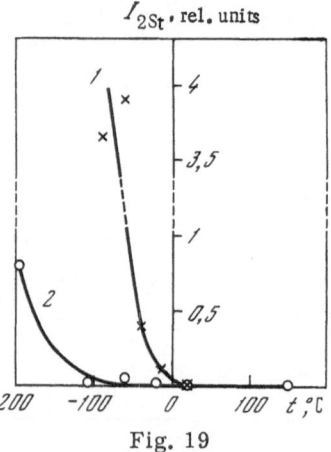

Fig. 18   Fig. 19

Fig. 18. Temperature dependences of the intensity of the first SRS Stokes components in calcite (1) and in carbon disulfide (2) and of the intensities of the ordinary Raman scattering oscillation at 1085 cm^{-1} in calcite (dash–dot line) and 656 cm^{-1} in carbon disulfide (dashed line).

Fig. 19. Temperature dependences of the intensities of the second Stokes component of SRS in carbon disulfide (1) and in calcite (2). The intensity at +20°C is taken to be 100.

components. Figure 19 shows plots of the temperature dependence of the intensity of the second Stokes component of SRS in carbon disulfide and in calcite, in relative units. In the temperature interval from +20°C to −100°C the intensity of the second Stokes component of the SRS in carbon disulfide increases by 2.5 orders, and in calcite by 1.5 order.

Thus, the temperature variations of the coefficient of conversion of the exciting radiation into SRS are determined to a considerable degree by the change of the probability of the ordinary Raman scattering with temperatures. This process alone, however, does not explain the temperature dependence of the intensity of the SRS in carbon disulfide. It appears that this dependence is connected to a considerable degree with the change of the regime of the self-focusing of the exciting radiation (see Fig. 15). The role of the self-focusing of the exciting radiation in the onset of the SRS was confirmed by the experiments of Rivoire and Beaudoin [84, 85], who observed an influence of the Kerr constant of the medium on the temperature dependence of the SRS threshold.

2. Dependence of the Coefficient of Conversion of the Exciting Radiation into SRS on the Thickness of the Scattering Layer

We measured in liquid nitrogen the dependence of the coefficient of conversion of the exciting radiation into SRS on the thickness of the scattering layer. The energies of the first Stokes component of the SRS in the forward and backward directions were measured with the aid of an FÉK-15 coaxial photocell. The results of the measurement of the coefficient of conversion η at various thickness of the scattering layer are given in Table 9.

As seen from the table, the coefficient of conversion into the first Stokes component in the forward direction increased from 0.5 to 12% when the thickness of the scattering layer increased from 8 to 20 mm, and with further increase of the thickness to 100 mm the coefficient remained practically unchanged. In the backward direction, the maximum energy was converted into the first SRS Stokes component in a scattering layer

TABLE 9

Direction	8 mm	20 mm	50 mm	100 mm
Forward	0,5	12	12	12
Backward	0,3	2,8	4	2,4

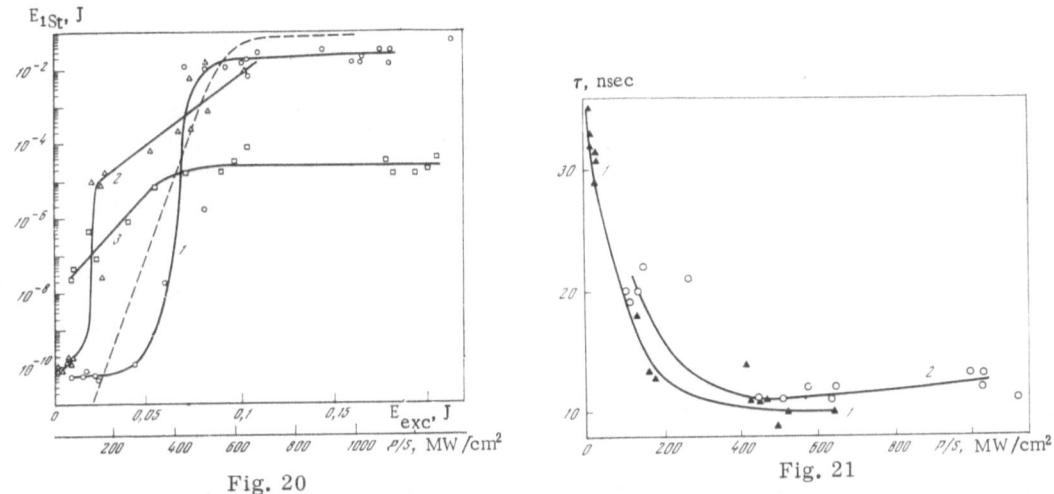

Fig. 20

Fig. 21

Fig. 20. Dependence of the energy of the first SRS Stokes component in calcite (1), nitrobenzene (2), and carbon disulfide (3) on the exciting-radiation power density. The solid curves were drawn through the experimental points, while the dashed line was calculated for calcite from formula (10).

Fig. 21. Dependence of the duration of the pulse of the first SRS Stokes component in nitrobenzene (1) and calcite (2) on the exciting-radiation power density.

50 mm thick. The change of the energy of the first SRS Stokes component with increasing length of the cell was investigated in [14] (in benzene, in the forward direction) and in [45] (in toluene, in the forward and backward directions). In all cases, the energy of the first SRS Stokes component first increased and then reached saturation.

3. Dependence of the Coefficient of Conversion into SRS on the Exciting-Radiation Power Density

Investigations of the dependence of the energy of the first SRS Stokes component on the exciting-radiation energy were carried out by many workers in liquid nitrogen [33], in acetone [18], in calcite [76], and in powdered stilbene [78].

We have performed similar measurements on calcite [86], nitrobenzene, and carbon sulfide under the conditions of our experiment. The SRS energy was measured with an ÉLU-FT receiver, which had high sensitivity and made it possible to measure SRS energies down to 10^{-11} J. The energy of the exciting radiation ranged from 0.01 to 0.02 J. Figure 20 shows the dependences of the energy of the first SRS Stokes component in calcite, nitrobenzene, and carbon disulfide on the excitation power density. The abscissas show also the values of the excitation energy. The character of the curves for the investigated substances is approximately the same. Each curve consists of several regions: a region of slow growth of the SRS energy, a region of fast growth where the SRS energy changes by several orders of magnitude for a small change of the energy of the exciting radiation, and a saturation region. In carbon disulfide, the abrupt growth of the energy of the first SRS Stokes component takes place at a low exciting-radiation power density and the values of the SRS energy in the lower part of the region of the rapid energy variation were too small to be registered. It can only be stated that at an exciting-radiation power density 40 MW/cm^2 the energy of the first SRS Stokes component is less than 10^{-10} J. In calcite, the energy of the first SRS Stokes component increases by seven orders of magnitude when the exciting-radiation power density increases from 350 to 500 MW/cm^2. We observed also a region where the energy of the first Stokes component of the SRS in liquid nitrogen increased abruptly following a negligible change in the energy of the exciting radiation in the region of the SRS saturation [67].

It should be noted that the SRS "energy" curves obtained by various workers under significantly different experimental conditions, with a different mode composition of the exciting radiation, and in scattering media with different optical characteristics, are all similar.

Plots of the duration of the pulse of the first SRS Stokes component (τ) in calcite and in nitrobenzene against the exciting-radiation power density are shown in Fig. 21. At low energies, the duration of the pulse is maximal and is close to the duration of the exciting-radiation pulse. In the region where the energy of the first SRS Stokes component has an abrupt growth, an appreciable decrease of the pulse duration of this component is observed.

We have compared the experimental dependence of the SRS energy as obtained by us on the energy of the exciting radiation with the theoretical value for single-crystal calcite at liquid-nitrogen temperature [86]. We observed in that case no self-focusing of the exciting radiation, and there was no feedback due to SMBS (inasmuch as no SMBS was excited at −196°C), nor was SRS generated (the crystal was homogeneous and the angle between the entrance and exit faces was 10°). Thus, the experimental conditions were closest to those for which the theoretical relation was obtained. The connection between the SRS intensity and the intensity of the exciting radiation was investigated theoretically by many workers [5, 11, 34]. We have obtained theoretical estimates in accordance with formula (10) [34]. For comparison with the experimental data, we plotted the values of the energy of the first SRS Stokes component

$$E_s = I_s \tau, \tag{26}$$

where τ is the duration of the SRS pulse.

The theoretical plot of the energy of the first SRS Stokes component in calcite vs. the energy of the exciting radiation is shown in Fig. 20. It is seen from the figure that the experimental curve does not agree with the theoretical one. The experimentally obtained dependence differs substantially from an exponential one. The theoretically calculated gain can be obtained from the slope of the curve. The calculated result corresponds to the slope of the curve in a region 300-400 MW/cm^2, and at higher values of the exciting-radiation power density the experimental value of the gain greatly exceeds the theoretical one.

The cause of the rapid growth of the first SRS Stokes component in a narrow interval of the exciting-radiation energies is presently under discussion. It was proposed that the jump is due to the generation. The feedback can be due to reflection of light from the cell windows, from the end faces of the crystal [76], or from inhomogeneities of the medium, or else to scattering in the backward direction (SMBS or SRS) [33]. In our investigations the windows of the dewar or the faces of the crystal were inclined 10°, but nevertheless the jump was observed and its magnitude in calcite (seven orders) was the same as in [76], where the crystal faces were parallel. A rapid growth of the energy was observed also in liquid nitrogen and calcite cooled to the temperature of liquid nitrogen, where there is no SMBS. The generation due to the feedback caused by the SRS in the backward direction apparently plays likewise no substantial role, for if generation were to occur the first SRS Stokes component would have the same energy in the forward and backward directions. The largest energy growth was observed in the calcite single crystal, the optically most homogeneous substance of all the investigated ones.

It was shown in [18] that an important role in the onset of the jump can be played by the self-focusing of the light. However, in calcite we did not observe any self-focusing of the exciting radiation, while in liquid nitrogen the self-focusing was observed at a higher exciting-radiation power density than in the region of the rapid growth of the energy of the first SRS Stokes component.

4. Investigation of the Dependence of the SRS Self-Focusing

Regime on the SRS Power

Simultaneously with the measurement of the coefficient of conversion of the exciting radiation into SRS under various excitation conditions, we investigated the self-focusing regime. A comparison of the results has shown that the SRS self-focusing regime is determined to a considerable degree by the power transformed into the first Stokes component of the SRS. Figure 22 shows plots of the number and diameter of the self-focusing dots of the first SRS Stokes component in liquid nitrogen and in calcite against the power of the first SRS Stokes component in these substances. The number and the diameter of the dots in the beam cross section were measured for nitrogen near the entrance window of the dewar, and for calcite near the entrance and exit faces of the crystal. The data for liquid nitrogen shown in the figure were obtained in an investigation of the distribution of the intensity in the field of the first SRS Stokes component as a function of the thickness of the scattering layer. The largest number of dots and their smallest diameter were observed in a vessel 50 mm thick, where the conversion of the energy into the first SRS Stokes component in the backward direction is largest. The numbers and diameters of the self-focusing dots obtained when the energy of the exciting radiation was varied lie approximately on these same curves. The plots for calcite, shown in Fig. 22, were obtained in a simultaneous investigation of the self-focusing regime, of the energy, and of the duration of the first SRS Stokes component with variation of the exciting-radiation power density. As seen from the figure, with increasing power of the first

SRS Stokes component the number of dots increase and their diameter decreases. It should be noted that the dots obtained in investigations of the self-focusing regime in the cross section of the beam of the first Stokes component near the entrance and exit faces of the crystal lie on these same curves.

A similar dependence of the self-focusing regime on the radiation power was observed for nitrobenzene. Plots of this dependence are shown in Fig. 23. This substance is characterized by the fact that when the first Stokes component is considerably changed only one dot is observed in the SRS field. When a power 10^6 W is reached, the number of dots increases to 10, and the diameter decreases to 20 μm. Near the entrance window of the cell (in the backward direction) the number of dots was larger than near the exit, and the diameters were smaller. However, an abrupt increase in the number of dots was observed at the same power of the first SRS Stokes component.

In media with large Kerr constants, in contrast to liquid nitrogen and calcite, similar changes in the self-focusing regime were observed also for the exciting radiation. A decrease of the diameter of the self-focusing dots of the exciting radiation with increase of power of the latter was observed for CS_2 [62]. We observed a change in the picture of the self-focusing of the exciting radiation when its power was changed in nitrobenzene. Figure 24 shows plots of the numbers and diameters of the self-focusing dots of the exciting radiation on its power in nitrobenzene. A considerable growth of the number of dots and the decrease of their diameter were observed at a power 10^6 W. At an SRS power higher than 10^6 W, the distribution of the intensity in the field of the first Stokes component near the exit face of the sample became diffuse. The change of SRS self-focusing regime with temperature, which we observed in carbon disulfide and calcite (see Figs. 13 and 15), is also due with the change in the power of the first SRS Stokes component. The maximum number and the minimal diameter of the dots was observed in carbon disulfide at temperatures from -40 to $-70°$C, and in calcite at $-196°$C. At the same temperatures, the coefficient of conversion of the exciting radiation into SRS is maximal (see Table 6). In carbon disulfide, cooling the medium from -70 to $-90°$C decreases the coefficient of conversion of the exciting radiation into SRS and the power of the first Stokes component, and simultaneously changes the self-focusing regime (decreases the number of dots and increases their diameter). Our investigations show that the numbers and diameters of the dots in self-focusing of the first Stokes component of SRS of light are determined by the power of the radiation converted into this component. At a power less than 10^5 W, the dots are larger (large-scale self-focusing), and then when the power is increased to $5 \cdot 10^5$-10^6 W the thick filaments break up into thinner ones, and small-scale self-focusing appears.

5. Radiation Power Density at the SRS Self-Focusing Dots

Our results have made it possible to determine the radiation power density at the dots produced by the self-focusing of the SRS of light. The power density of the focused radiation equals P/S, where P is the radiation power and $S = N\pi r^2$ is the total cross-sectional area of the focused beam (N is the number of dots and r is the average radius of the self-focusing dot). Table 10 gives the number N, the radius r, and the total area S of the self-focusing dots of the first SRS Stokes component in calcite.

As seen from the table, the area remains constant when the excitation energy varies from 0.08 to 0.185 J, with accuracy 4%. Measurements of the cross-sectional areas of the focused beams in other substances have shown that the area does not depend on the excitation conditions and is a constant for a given substance.

The radiation power density in the dots increases when their diameter decreases and when the power increases, up to a certain value, beyond which it becomes constant. The maximum power density of the focused beam is reached for each substance at a definite value of the filament diameter and is a quantity that characterizes the given substance. Table 11 gives the values of the total area of the self-focusing dots and of the power density of the first SRS Stokes component in calcite, liquid nitrogen, nitrobenzene, and carbon disulfide. The error in the measurement of the areas was 10%. As seen from the table, the maximum power density is observed in nitrobenzene and in carbon disulfide (10^5 MW/cm^2), and the minimal one in calcite ($6 \cdot 10^3$ MW/cm^2).

TABLE 10

E_{exc}, J	N	$r \cdot 10^3$, cm	$S \cdot 10^4$, cm^2	E_{exc}, J	N	$r \cdot 10^3$, cm	$S \cdot 10^4$, cm^2
0.08	5	4.5	3.2	0.17	17	2.5	3.3
0.1	5	4.2	2.8	0.18	28	1.9	3.2
0.13	8	3.5	3.1	0.185	90	1.0	3.1
0.15	11	3.1	3.3				

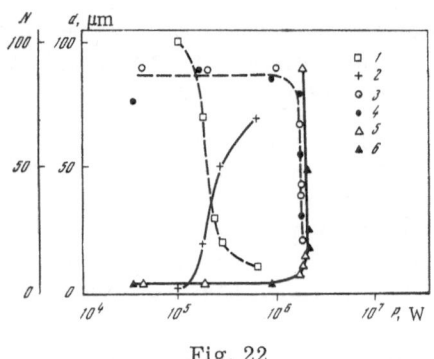
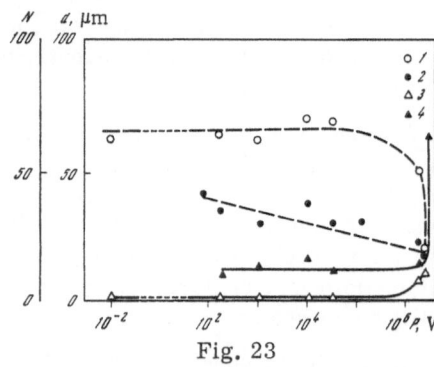

Fig. 22   Fig. 23

Fig. 22. Dependence of the number and the diameter of the dots of the SRS self-focusing on the power of the first Stokes component in liquid nitrogen and in calcite.
(1,2) Diameter and number of dots in liquid nitrogen; (3, 4) diameter of dots in calcite in the forward and backward directions; (5,6) number of dots in calcite in the
forward and backward directions.

Fig. 23. Dependence of the number and diameter of the SRS self-focusing dots on
the power of the first Stokes component in nitrobenzene. (1, 2) Diameter of dots
forward and backward; (3, 4) number of dots forward and backward.

6. Discussion of the Results of an Investigation of the Self-Focusing in Substances with Different Kerr Constants

Our experimental results have shown that in contrast to the prevailing notions, self-focusing of light
occurs in substances with small Kerr constants when excited by nanosecond pulses. In calcite we have observed self-focusing of the first SRS Stokes component, whereas the distribution of the intensity in the exciting-radiation beam cross section was diffuse. In liquid nitrogen the self-focusing threshold of the SRS was
much lower than the self-focusing threshold of the exciting radiation. In substances with large Kerr constant,
depending on the energy converted into SRS, we observed different versions of the distribution of the self-focusing dots on the end faces of the cell. The pictures of the self-focusing of the SRS and of the exciting radiation could be the same or different; sometimes the self-focusing dots were produced only in the SRS beam cross
section. The number and diameter of the self-focusing dots of the SRS depended on the coefficient of conversion
of the exciting radiation into SRS.

The results indicate that the development of self-focusing of SRS of light in the investigated substances is
autonomous and independent of the self-focusing of the exciting radiation. Similar results were later obtained
by Korobkin, Lugovoi, Prokhorov, and Serov [87]. They used a single-mode laser. They investigated the kinetics of the self-focusing of the SRS and of the exciting radiation with high time resolution (10^{-11} sec) in nitrobenzene and carbon disulfide. They observed a secondary (compared with the self-focusing of the exciting radiation) development of self-focusing of the SRS pulse. The power of the SRS radiation produced in the foci of
the excitation radiation was higher than critical, and the SRS pulse underwent nonlinear refraction, producing
in turn a multifocus structure. The SRS were observed on the screen of the image converter only after the passage of the first focus of the exciting radiation. The picture of the distribution of the SRS self-focusing dots did
not always duplicate the distribution of the self-focusing dots of the exciting radiation.

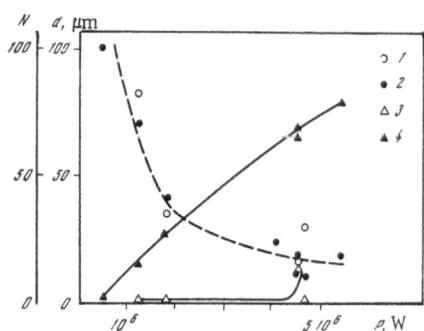

Fig. 24. Dependence of the number and
diameter of the self-focusing dots of the
undisplaced component and of the SMBS
on the exciting-radiation power in nitrobenzene. (1, 2) Diameter of forward and
backward dots; (3, 4) number of forward
and backward dots.

TABLE 11

Parameter	CaCO$_3$	N$_2$	C$_6$H$_5$NO$_2$	CS$_2$
S, cm^2	$3 \cdot 10^{-4}$	$1.5 \cdot 10^{-4}$	$3.3 \cdot 10^{-5}$	$2 \cdot 10^{-4}$
P/S, MW/cm^2	$6 \cdot 10^3$	10^4	$8 \cdot 10^4$	$9 \cdot 10^4$

TABLE 12

Substance	t, °C	$n_2 \cdot 10^{13}$, cm^3/erg	P_{theor}, kW	P_{exp}, kW — our data	P_{exp}, kW — results of others
CS$_2$	+20	120 [88, 55] [58] 310 [57] 15 [56]	6 2,3 47	200	25 [63] 15 [89]
	−30	218 [88]	3.2	—	—
C$_6$H$_5$NO$_2$	+20	152 [55, 58]	5	500	19 [66]
N$_2$	−196	3.2 [88, 57]	400	720	—

The refractive-index change that leads to nonlinear refraction of the light can be the result of various processes that occur in the medium under the influence of the laser-radiation field. The principal mechanism that leads to a change in the refractive index of the medium and to self-focusing of light is the orientation of the molecules by the high-frequency light field due to the Kerr effect. The threshold light self-focusing power connected with the Kerr effect can be estimated from formula (25).

Although theoretical estimates of the threshold power were made many times, an analysis of the data contained in the literature has shown that they are quite contradictory. Different authors obtained significantly different experimental values of the Kerr constant, and this led to discrepancies in the calculation of the non-linear refractive index. The handbook [88] gives the Kerr constant obtained for a static field. Shen has shown theoretically [58], and Mayer and Gires have shown experimentally [55], that the electric field of a light beam from a laser produces effects similar to that action which is exerted by a static field on molecules without a constant moment. The high-frequency Kerr constants obtained by these authors for CS$_2$ are close to the Kerr constant in a constant field. The same should be observed also for liquid nitrogen. For nitrobenzene, the Kerr constant in the field of laser radiation is not equal to its value in a static field. An experimental value of the high-frequency Kerr constant of nitrobenzene is given in [55], and a calculated value is given in [58]. These two values are close. One other method of determining the nonlinear refractive index is contained in the papers of Maker, Terhune, and Savage [56] and of McWane and Sealer [57]. They measured the rotation of the oscillation ellipse of elliptically polarized light after passage of the light through the medium. This rotation is expressed in terms of the field components E_x and E_y in the following manner:

$$ a = \frac{\pi \omega}{cn} 2BE_x E_y z, $$

where α is the rotation of the oscillation ellipse, c is the speed of light in vacuum, n is the refractive index, and z is the thickness of the layer of the nonlinear medium.

The values of the Kerr constant obtained in these papers differ by approximately 20 times. McWane and Sealer attribute this difference to the fact that they used unfocused laser radiation, whereas Maker, Terhune, and Savage focused the laser beam into the medium.

Table 12 lists the values of n_2 and the exciting-radiation self-focusing threshold powers, calculated theoretically (P_{theor}) from formula (25) and obtained experimentally (P_{exp}) for carbon disulfide, nitrobenzene, and liquid nitrogen.

It is seen from the table that the threshold of self-focusing of the exciting radiation is higher in our experiments than its theoretically calculated values. This may be due to the difference between the intensity distribution in the cross section of the real beam and the ideal distribution. Under real conditions, not the entire beam is self-focused, only part of it. This effect was most clearly observed in nitrobenzene, where the intensity of the diffuse background, not focused into dots, was high (see Fig. 16). A beam having a real divergence can become stratified and individual parts of it can become self-channeled. In this case only a certain fraction of the energy of the exciting radiation is concentrated in the channel, and the threshold power for the production of such a channel exceeds the critical value by one or two orders [53]. Estimates made by us for the conditions of our experiments have shown that the threshold power should exceed the critical one by 20 times, which is approximately by how much our experimental results differ from the calculated threshold power. Thus, in nitrobenzene and in carbon disulfide, the main contribution to the **refractive-index change** that causes the self-focusing is made by the Kerr effect.

The threshold self-focusing power of the first SRS Stokes component in liquid nitrogen, calculated from formula (25), amounts to 670 kW. We have observed self-focusing of SRS in liquid nitrogen at a first SRS Stokes component of 30 kW. We observed also self-focusing of SRS in calcite, although no Kerr effect should occur in this substance. This allows us to assume that the self-focusing of SRS in substances with small Kerr constant is due to another mechanism of change in the refractive index.

In single-crystal calcite, where the molecular rotations are hindered and the Kerr constant is quite small, the change of the refractive index may be due to electrostriction. Electrostriction leads as a rule to much smaller changes of the refractive index than the Kerr effect. The electrostriction coefficients of substances have a strong positive temperature dependence. Thus, the role of the electrostriction in the onset of self-focusing can be determined from the temperature dependence of this effect.

An increase in the calcite temperature should lead not only to an increase of the electrostriction coefficient, but also to a decrease of the coefficient of conversion into SRS, and consequently to an increase in the radiation energy at the undisplaced frequency. The result would be a decrease in the threshold of the self-focusing of the exciting radiation when the calcite is heated. Nonetheless, no self-focusing of the exciting radiation was observed when the crystal temperature was changed from -196 to $+20°C$. It appears that electrostriction does not play a significant role in self-focusing in calcite.

The nonlinearity of the refractive index of a medium can result from rocking of the molecule in the field of neighboring molecules [90]. This model was first proposed by Starunov to interpret the far wing of the Rayleigh line [91]. This mechanism plays a role mainly when ultrashort pulses pass through the medium, whereas in the case of nanosecond pulses it goes over into the Kerr effect.

It is not excluded that the increase of the refractive index, which causes the self-focusing of SRS in substances with small Kerr constants, is due to the increase of the polarizability of the molecules upon excitation. The first indication of the possibility of this mechanism is contained in the papers of Townes and co-workers [64] and of Askar'yan [92]. The change of the polarizability of the molecules under the influence of the nonlinear processes that occur in the substance was investigated in detail by Butylkin, Kaplan, and Khronopulo [52]. When SRS is excited in a medium, the conversion of the pump into the Stokes component occurs over a short length (for example over a length $L = 0.1$ cm for benzene). Thus, the layer in which the susceptibility of the molecule differs noticeably from the equilibrium value can be regarded as a thin lens made up by the pump waves and the Stokes component with a still-undistorted geometry. The calculation of such a lens shows that if the input intensity has a quadratic distribution over the cross section of the beam

$$E_{In}^2 = E_0^2 (1 - \rho^2/a^2), \tag{27}$$

where E_0^2 is the intensity at the center of the beam and a is the radius of the beam, the focal length is equal to

$$F = \frac{k_0 a^2 \hbar}{2\tau (\varkappa_2 - \varkappa_1)}, \tag{28}$$

where $k_0 = 2\pi/\lambda_0$; τ is the relaxation time of the vibrational level; $\varkappa_2$ and $\varkappa_1$ are the polarizabilities in the excited and unexcited states.

The thin-lens approximation is valid at $F > L$, and leads to the condition for the entry radius of the beam

$$a > a_0 = \left(\frac{2\tau (\varkappa_2 - \varkappa_1) L}{k_0 \hbar} \right)^{1/2}.$$

For benzene $a \approx 10^{-4}$ cm, and for compressed hydrogen 0.1 cm. The threshold pump intensity needed for self-focusing of SRS is obtained by equating F to the diffraction length of the beam $R_g = k_0 a^2$:

$$|E_{thr}|^2 = \frac{\hbar}{2\tau (\varkappa_2 - \varkappa_1)} = \mathscr{E}^2. \tag{29}$$

Allowance for the finite SRS threshold (E_{SRS}^2) leads to the following value: $E_{thr}^2 = \xi (2E_{SRS}^2 + \mathscr{E}^2)$, where $\xi = 1$ at $E_{SRS}^2 \ll \mathscr{E}^2$ and $\xi = 1/2$ at $E_{SRS}^2 \gtrsim \mathscr{E}^2$.

For self-focusing of SRS the threshold quantity is not the total power flux in the beam, but the flux density. If the input pump intensity exceeds by many times the SRS threshold, then we can expect the appearance of a number of lenses connected with successive transformation of the radiation into Stokes components of higher order.

In the calculation of the threshold of this effect in [52], τ was assumed equal to $3 \cdot 10^{-8}$ sec, $\varkappa_2 - \varkappa_1 \approx 10^{-25}$. For our case the threshold power of the exciting radiation is then 20 MW/cm^2. The threshold of the self-focusing due to the Kerr effect is smaller than this value in substances with large Kerr constants, and larger in substances with small Kerr constants. Thus, this effect can play an important role in self-focusing of light in substances with small Kerr constants.

Theoretical estimate of the change of the refractive index under the influence of the excitation of the vibrations of the molecules in SRS were made also by Vilgel'mi and Goiman [93]. They calculated the change of the polarizability upon excitation, using an adiabatic approximation method based on Placzek's polarizability theory. It turned out that the relative change of the polarizability can amount to several percent (up to 19%), i.e., it is comparable with its change under the influence of other effects, such as the Kerr effect.

They have assumed that at the proposed high intensities and in sufficiently long cells the coefficient of conversion of the exciting-radiation photons into Stokes photons reaches values on the order of unity, and the process of generation of Stokes photons occurs over a rather short length (L $\lesssim$ 10 mm). In substances with large Kerr constants, such as CS_2, under ordinary conditions, the change of the refractive index as a result of the SRS of the light is negligible compared with the change due to the Kerr effect. On the contrary, for a substance with a low Kerr constant, such as liquid nitrogen, the excitation of the molecule vibrations can make a substantial contribution to the change in the refractive index.

The presented estimates of the threshold power density and of the change of the refractive index show that in our case, in substances with small Kerr constant, it is possible to observe SRS self-focusing due to the increase of the polarizability of the molecules upon excitation. However, to calculate exactly the contribution made by this mechanism to the SRS self-focusing we need more accurate theoretical calculations and possibly also additional experiments.

The self-focusing of the exciting radiation may not occur, due to an appreciable conversion into SRS, which leads to a decrease in the power of the laser light. This is particularly significant for self-focusing of SRS, since the maximum change of the refractive index occurs in this case in those regions where the conversion into SRS is largest and consequently the exciting radiation is strongly attenuated. In addition, it should be noted that, according to [87], the SRS pulse is produced after the passage of the maximum intensity of the exciting-light pulse.

An important role in the onset of self-focusing can be played by the fact that the beam of the exciting radiation is not homogeneous in space. Akhmanov, Sukhorukov, and Khokhlov [94] have shown that a Gaussian beam exhibits a tendency to stratification as it propagates in a nonlinear medium. Small deviations from a Gaussian shape lead to the onset of transverse inhomogeneity of the beam even if it is close to being axially symmetrical [95]. Theoretical calculations by Bespalov and Talanov [96] have shown that in a nonlinear medium an inhomogeneous beam with a power greatly exceeding the critical value is unstable and breaks up into individual beams with power on the order of the critical value. The intensities of the beams that are separated from the smooth envelope of the laser pulse can increase rapidly in a nonlinear medium. The SRS appears at the locations of the inhomogeneity (maximal intensity) of the beam of the exciting radiation. The nonlinearity of the conversion of the energy into SRS can cause this structure to become enhanced in the field of the Stokes component of the SRS and concentration of the energy in certain regions of the beam. This in turn can lower the threshold power of the self-focusing of the SRS and increase the number of self-focusing dots. It appears that this effect exerts an influence on the formation of the SRS beam during the initial stage of the self-focusing. However, if it were to play a decisive role, then the dimension of the observed points in the SRS field should not exceed the dimension of the inhomogeneities in the field of the exciting radiation. We, however, observed, near the self-focusing threshold, dots that exceed the dimensions of the inhomogeneities by almost one order of magnitude. With increasing coefficient of conversion into SRS, the diameter of the dots decreased to 10 μm. So small a size of the dots also indicates that their cause is self-focusing of the light.

The procedure used in a number of experimental investigations on which the main concepts concerning the development of self-focusing of light in a medium are based does not make it possible to assess the spectral composition of the radiation in the self-focusing "filaments." The filaments were observed by viewing the cell laterally and were revealed by the scattering of light from an auxiliary source by the inhomogeneities

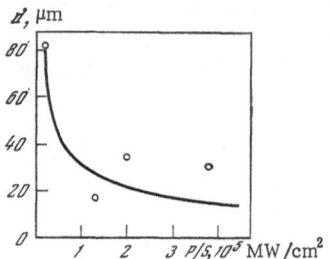

Fig. 25. Dependence of the diameter of the dots of self-focusing of the exciting light in nitrobenzene on the radiation power density. Solid curve — calculation; points — experiment.

caused by the change of the refractive index [48]. The self-focusing of the SRS could not be registered and was not regarded as an independent process. In this connection, the theoretical estimates of the diameters of the self-focusing regions and of the radiation power density were always carried out for laser light.

The decrease in the diameter and the increase in the number of self-focusing dots of the exciting radiation in nitrobenzene with increasing excitation power, observed in our study, agree qualitatively with the results obtained by others. Thus, a decrease in the diameter of the dots from dozens and hundreds of microns to several microns was observed in carbon disulfide [62], as well as an increase in the number [64], when the laser power was increased.

The smallest dot diameter obtained in our study is 10 μm. Other workers observed dots of 4-5 μm diameter. These exceed the theoretical values (calculations yield a minimum dot radius on the order of the radiation wavelength) [53, 94]. The most probable factor that hinders the formation of filaments with radius on the order of the wavelength is the attenuation of the intensity of the laser radiation as a result of the conversion into SRS.

As shown by our experiments, the minimum diameter of the self-focusing points and the maximum density of the power radiation in them are constants for a given substance. This agrees with the conclusion arrived at by Lugovoi and Prokhorov in a theoretical investigation of the structure and dimensions of the focal spots produced when laser radiation interacts with matter [26]. It is shown in this reference that the values of the maximum intensity $|E_{fs}|^2$ and of the diameter d_{fs} of the focal spot are connected by the relation [26]

$$|E_{fs}|^2 d_{fs}^2 \approx 0.18 \frac{\lambda^2}{n_2} . \tag{30}$$

The authors arrived at the conclusion that the energy density at the center of the focus and the diameter of the focus are determined only by the constants of the medium. From (30) we can obtain the conditions for the power density and the diameter of the self-focusing point:

$$(P/S) d^2 \approx 0.18 \frac{c n_0 \lambda^2}{8 \pi n_2} . \tag{31}$$

Although the power density and the diameter of the self-focusing points varied in our experiments with changing excitation conditions, this relation for the self-focusing of the exciting radiation was satisfied on the whole. This is illustrated in Fig. 25. It shows a plot of the diameter of the filaments against the radiation power density in filaments, as obtained from formula (31). The same figure shows the points obtained from our experiments. It is seen that the experimental data are close to the curve plotted in accordance with formula (31).

In carbon disulfide and in nitrobenzene, the radiation power density at the SRS self-focusing dots was of the same order as in the self-focusing dots of the exciting radiation. In calcite and liquid nitrogen the power density in the points of self-focusing of the SRS was approximately lower by one order of magnitude than in carbon disulfide and in nitrobenzene. The minimal diameter of the SRS self-focusing dots and the maximum radiation power density in them, just as in the case of self-focusing of laser light, were quantities characteristic of the given substance.

The SRS power flowing through the cross section of each self-focusing dot amounted in our experiments to 10^4-10^5 W. This agrees with theoretical estimates based on formula (31). If the lifetime of the "filaments" (or the time of passage of the traveling focus through the medium) is 10^{-10}-10^{-11} sec [26, 48], then the energy contained in each SRS self-focusing dot is several ergs. This agrees with the estimates of the energy of the exciting-radiation filaments in [48, 64].

The experimental results obtained in the present study lead to essentially new concepts concerning the development of self-focusing of light in media in which simulated Raman scattering of light takes place. It is seen from the presented examples that even in media with large Kerr constants, such as carbon disulfide, the self-focusing phenomenon is observed most clearly and in a variety of aspects precisely in SRS of light, and leads to a number of interesting effects, particularly to generation of ultrashort pulses. In media with low Kerr constants, as shown by our investigations, self-focusing of SRS can occur and is due to the change in the polarizability of the molecules upon excitation. For a more detailed elucidation of the kinetics of self-focusing of light in media, experiments must be performed that can trace this process in different spectral regions in space and in time. The need for spectral resolution of the light in investigations of this kind imposes very stringent requirements on the sensitivity of the apparatus used for the registration.

In conclusion, I am deeply grateful to my scientific adviser A. I. Sokolovskaya for direction and constant attention to the work, and also to M. M. Sushchinskii for interest in the work and for valuable comments.

LITERATURE CITED

1. E. J. Woodbury and W. K. Ng, Proc. IRE, 50:2367 (1962).
2. G. Eckhardt, R. W. Hellwarth, F. J. McClune, S. E. Schwarz, D. Weiner, and E. J. Woodbury, Phys. Rev. Lett., 9:455 (1962); Electron. Des., 11:28 (1963).
3. M. Geller, D. P. Bortfeld, and W. R. Sooy, Appl. Phys. Lett., 3:36 (1963).
4. V. A. Zubov, M. M. Sushchinskii, and I. K. Shuvalov, Usp. Fiz. Nauk, 83:197 (1964); 89:49 (1966).
5. P. A. Apanasevich, Zh. Prikl. Spektrosk., 6:183, 322 (1967).
6. E. Garmire, F. Pandarese, and C. H. Townes, Phys. Rev. Lett., 11:160 (1963).
7. R. Loudon, Proc. Phys. Soc. London Sect. A, 82:393 (1963); Proc. R. Soc. London Ser. A, 275:218 (1963).
8. R. W. Terhune, Solid State Des., 4:38 (1963).
9. N. Bloembergen and Y. R. Shen, Phys. Rev., 133:A37 (1964).
10. V. T. Platonenko and R. V. Khokhlov, Zh. Éksp. Teor. Fiz., 46:555 (1964).
11. R. W. Hellwarth, Phys. Rev., 130:1850 (1963).
12. H. J. Zeiger, P. E. Tannenwald, S. Kern, and R. Herendeen, Phys. Rev. Lett., 11:419 (1963).
13. F. J. McClung, W. G. Wagner, and D. Weiner, Phys. Rev. Lett., 15:96 (1965).
14. P. Lallemand and N. Bloembergen, Appl. Phys. Lett., 6:210, 212 (1965); Phys. Rev. Lett., 15:1010 (1965).
15. C. C. Wang and G. W. Racette, IEEE J. Quantum Electron., QE-2:65 (1966).
16. B. P. Stoicheff, Phys. Lett., 7:186 (1963).
17. P. D. Maker and R. W. Terhune, Phys. Rev., 137:A801 (1965).
18. G. G. Bret and M. M. Denariez, Appl. Phys. Lett., 8:151 (1966).
19. E. Garmire, Phys. Lett., 17:251 (1965).
20. G. Rivoire, J. Phys. (Paris), 28:711 (1967).
21. K. Shimoda, Jpn. J. Appl. Phys., 5:86 (1966).
22. G. A. Askar'yan, Zh. Éksp. Teor. Fiz., 42:1567 (1962).
23. N. F. Pilipetskii and A. R. Rustamov, Pis'ma Zh. Éksp. Teor. Fiz., 2:88 (1965).
24. R. Y. Chiao, E. Garmire, and C. H. Townes, Phys. Rev. Lett., 13:479 (1964).
25. A. L. Dyshko, V. N. Lugovoi, and A. M. Prokhorov, Pis'ma Zh. Éksp. Teor. Fiz., 6:655 (1967).
26. V. N. Lugovoi and A. M. Prokhorov, Usp. Fiz. Nauk, 111:203 (1973).
27. V. V. Korobkin, A. M. Prokhorov, R. V. Serov, and M. Ya. Shchelev, Pis'ma Zh. Éksp. Teor. Fiz., 11:153 (1970); V. V. Korobkin, Usp. Fiz. Nauk, 107:512 (1972).
28. Y. R. Shen and Y. Shaham, Phys. Rev. Lett., 15:1008 (1965).
29. M. Maier and W. Kaiser, Phys. Lett., 21:529 (1966).
30. G. Placzek, The Rayleigh and Raman Scattering, US AEC Report UCRL-Trans-526(L) (1962).
31. S. A. Akhmanov and R. V. Khokhlov, Problems of Nonlinear Optics [in Russian], VINITI, Moscow (1964).
32. N. Bloembergen, Nonlinear Optics, Benjamin (1965).
33. J. B. Grun, A. K. McQuillan, and B. P. Stoicheff, Phys. Rev., 180:61 (1969).
34. M. M. Sushchinskii, Raman Spectra of Molecules and Crystals [in Russian], Nauka, Moscow (1969).
35. V. N. Lugovoi, Introduction to the Theory of Stimulated Raman Scattering [in Russian], Nauka, Moscow (1968).
36. A. D. Kudryavtseva, A. I. Sokolovskaya, and M. M. Sushchinskii, Zh. Éksp. Teor. Fiz., 59:1556 (1970).
37. A. I. Sokolovskaya, A. D. Kudryavtseva, and M. M. Sushchinskii, Nonlinear Optics – Proceedings of the 2nd All-Union Symposium on Nonlinear Optics [in Russian], Novosibirsk (1968), p. 277.

38. A. I. Sokolovskaya, A. D. Kudryavtseva, T. P. Zhbanova, and M. M. Sushchinskii, Zh. Éksp. Teor. Fiz., 53:429 (1967).

39. A. I. Sokolovskaya, A. D. Kudryavtseva, G. L. Brekhovskikh, and M. M. Sushchinskii, Zh. Éksp. Teor. Fiz., 57:1160 (1969).

40. A. I. Sokolovskaya, A. D. Kudryavtseva, G. L. Brekhovskikh, and M. M. Sushchinskii, Preprint FIAN, No. 2 (1969).

41. W. Heinicke and G. Winterling, Appl. Phys. Lett., 11:231 (1967).

42. A. I. Sokolovskaya, Tr. Fiz. Inst. Akad. Nauk SSSR, 27:63 (1964).

43. A. D. Kudryaetseva, A. I. Sokolovskaya, and M. M. Sushchinskii, Kratk. Soobshch. Fiz., No. 2:32 (1971).

44. A. D. Kudryavtseva, A. I. Sokolovskaya, and M. M. Sushchinskii, Laser und ihre Anwendungen, Vol. 15, VEB Deutscher Verlag der Wissenschaften, Berlin (1970), p. 955.

45. Y. R. Shen and Y. I. Shaham, Phys. Rev., 163:224 (1967).

46. W. H. Culver, J. T. A. Vanderslice, and V. M. Townsend, Appl. Phys. Lett., 12:189 (1968).

47. M. Maier, W. Kaiser, and J. A. Giordmaine, Phys. Rev. Lett., 17:1275 (1966).

48. F. Shimizu and B. P. Stoicheff, IEEE J. Quantum Electron., QE-5:544 (1969).

49. D. von der Linde, M. Maier, and W. Kaiser, Phys. Rev., 178:11 (1969).

50. M. A. Bol'shov, Yu. D. Golyaev, V. S. Dneprovskii, and I. I. Nurminskii, Zh. Éksp. Teor. Fiz., 57:346 (1969).

51. A. L. Golger, Vestn. Mosk.Gos. Univ., Ser. Fiz. Astron., 11:693 (1970).

52. V. S. Butylkin, A. E. Kaplan, and Yu. G. Khronopulo, Zh. Éksp. Teor. Fiz., 59:921 (1970).

53. B. Ya. Zel'dovich, V. I. Popovichev, V. V. Ragul'skii, and F. S. Faizullov, Pis'ma Zh. Éksp. Teor. Fiz., 15:160 (1972); O. Yu. Nosach, V. I. Popovichev, V. V. Ragul'skii, and F. S. Faizullov, Pis'ma Zh. Éksp. Teor. Fiz., 16:617 (1972).

54. P. L. Kelley, Phys. Rev. Lett., 15:1005 (1965).

55. G. Mayer and F. Gires, C. R. Acad. Sci., 258:2039 (1964).

56. P. D. Maker, R. W. Terhune, and C. M. Savage, Phys. Rev. Lett., 12:507 (1964).

57. P. D. McWane and D. A. Sealer, Appl. Phys. Lett., 8:278 (1966).

58. Y. R. Shen, Phys. Lett., 20:378 (1966).

59. V. I. Talanov, Pis'ma Zh. Éksp. Teor. Fiz., 2:218 (1965).

60. V. N. Lugovoi, Dokl. Akad. Nauk SSSR, 176:58 (1967).

61. M. M. T. Loy and Y. R. Shen, Appl. Phys. Lett., 19:285 (1971).

62. M. Maier, G. Wendl, and W. Kaiser, Phys. Rev. Lett., 24:352 (1970).

63. E. Garmire, R. Y. Chiao, and C. H. Townes, Phys. Rev. Lett., 16:347 (1966).

64. R. Y. Chiao, M. A. Johnson, S. Krinsky, H. A. Smith, C. H. Townes, and E. Garmire, IEEE J. Quantum Electron., QE-2:467 (1966).

65. N. Bloembergen, G. Bret, P. Lallemand, A. Pine, and P. Simova, IEEE J. Quantum Electron., QE-3:197 (1967).

66. C. C. Wang, Phys. Rev. Lett., 16:344 (1966).

67. A. D. Kudryavtseva, A. I. Sokolovskaya, and M. M. Sushchinskii, in: Kvantovaya Élektron. (Moscow), No. 7, 73 (1972).

68. A. I. Sokolovskaya, A. D. Kudryavtseva, and M. M. Sushchinskii, Nonlinear Processes in Optics, No. 2 — Proceedings of the 2nd Vavilov Conference on Nonlinear Optics [in Russian], Nauka, Novosibirsk (1972), p. 262.

69. E. A. Morozova, A. I. Sokolovskaya, A. D. Kudryavtseva, and M. M. Sushchinskii, in: Kvantovaya Élektron. (Moscow), No. 4, 76 (1973).

70. R. G. Brewer and C. H. Lee, Phys. Rev. Lett., 21:267 (1968).

71. A. J. Alcock, C. De Michelis, V. V. Korobkin, and M. C. Richardson, Appl. Phys. Lett., 14:145 (1969).

72. M. M. T. Loy and Y. R. Shen, Phys. Rev. Lett., 25:1333 (1970).

73. A. D. Kudryavtseva and A. I. Sokolovskaya, Kvantovaya Élektron. (Moscow), 1:964 (1974).

74. V. A. Zubov, A. V. Kraiskii, K. A. Prokhorov, M. M. Sushchinskii, and I. K. Shuvalov, Zh. Éksp. Teor. Fiz., 55:443 (1968).

75. F. G. Badalyan, M. E. Movsesyan, and Zh. O. Ninoyan, Nonlinear Optics — Proceedings of the 2nd All-Union Symposium on Nonlinear Optics [in Russian], Nauka, Novosibirsk (1968), p. 342.

76. G. Bisson and G. Mayer, J. Phys. (Paris), 29:97 (1968).

77. F. A. Korolev, Z. A. Baskakova, and V. I. Odintsov, Opt. Spektrosk., 28:1125 (1970).

78. G. V. Peregudov, E. N. Ragozin, and V. A. Chirkov, Zh. Éksp. Teor. Fiz., 63:421 (1972).

79. A. I. Sokolovskaya, A. D. Kudryavtseva, G. L. Brekhovskikh, and M. M. Sushchinskii, Preprint FIAN, No. 169 (1968).

80. A. I. Sokolovskaya and Z. Kecki, Bull. Acad. Pol. Sci., Ser. Sci. Chem. Geol. Geogr., 6:133 (1958).

81. G. P. Tulub and Ya. S. Bobovich, Opt. Spektrosk., 9:669 (1960).

82. K. Park, Phys. Lett., 22:39 (1966).

83. B. K. Narayanaswamy, Proc. Indian Acad. Sci. Sect. A, 26:511 (1947).

84. G. Rivoire and J.-L. Beaudoin, Onde Electr., 48:196 (1968).

85. G. Rivoire and J.-L. Beaudoin, J. Phys. (Paris), 29:759 (1968).

86. A. D. Kudryavtseva, E. A. Morozova, and M. M. Moiseenko, Kratk. Soobshch. Fiz., No. 10:31 (1973).

87. V. V. Korobkin, V. N. Lugovoi, A. M. Prokhorov, and R. V. Serov, Pis'ma Zh. Éksp. Teor. Fiz., 16:595 (1972).

88. Landolt-Börnstein Tables, Vol. II, Part 8.

89. C. C. Wang and G. W. Racette, Appl. Phys. Lett., 8:256 (1966).

90. R. Cubeddu, R. Polloni, C. A. Sacchi, and O. Svelto, Phys. Rev. A, 2:1955 (1970).

91. V. S. Starunov, Opt. Spektrosk., 18:300 (1965).

92. G. A. Askar'yan, Pis'ma Zh. Éksp. Teor. Fiz., 4:400 (1966).

93. B. Vil'gel'mi and É. Goiman, Zh. Prikl. Spektrosk., 19:550 (1973).

94. S. A. Akhmanov, A. P. Sukhorukov, and R. V. Khokhlov, Usp. Fiz. Nauk, 93:19 (1967).

95. V. N. Lugovoi and A. M. Prokhorov, Pis'ma Zh. Éksp. Teor. Fiz., 7:153 (1968).

96. V. I. Bespalov and V. I. Talanov, Pis'ma Zh. Éksp. Teor. Fiz., 3:471 (1966).

INVESTIGATION OF THE SPECTRAL DISTRIBUTION OF THE INTENSITY OF THE COMPONENTS OF STIMULATED RAMAN SCATTERING OF LIGHT IN CONDENSED MEDIA*

E. A. Morozova

Systematic investigations of the spectral distribution of the intensity of SRS components were performed for the first time. It is shown that the existing theoretical concepts do not describe the actually observed spectral distributions. The fine structure of the SRS lines was observed and investigated in detail. Different physical mechanisms of its appearance are discussed. The influence of self-focusing of SRS on the development of the spectral distributions of the intensity was investigated experimentally for the first time ever. It is shown that the appearance of large spectral broadenings at the SRS self-focusing dots and the angular directivity of the components in media with large Kerr constants can be explained from the point of view of the concept of the propagation of high-power optical radiation in a medium in the form of moving focal regions.

INTRODUCTION

The spectral composition of a medium is substantially altered when a powerful laser radiation propagates in it. This is caused by a large number of factors, including the onset of scatterings in the medium [stimulated Raman scattering (SRS), stimulated Mandelstam–Brillouin scattering (SMBS)] and the self-focusing of the light. In turn, the spectral composition of the SRS components depends on the optical constants and on the temperature of the medium, as well as on the spectral composition and intensity of the exciting radiation, and is closely connected with the development of other nonlinear effects in the medium simultaneously with the SRS. The present paper is devoted to an experimental investigation of the spectral distribution of the intensity of the SRS components following excitation by a laser that generates one longitudinal mode, in substances in the condensed state.

These investigations are of great interest from the point of view of clarifying the mechanism of the action of powerful laser radiation on molecules of a medium, the process of development of SRS of light in a medium, and the connection between this phenomenon and other nonlinear effects, primarily self-focusing of light. These investigations are of importance not only for the development of the theoretical understanding of the SRS phenomenon and the self-focusing of light, but also for practical applications. In particular, the results can be useful for the solution of an important problem in quantum electronics, namely the development of a coherent-radiation source with a controllable spectral makeup.

It is known that many problems in nonlinear optics, holography, quantum chemistry, and others require the use of radiation with very high degree of monochromaticity. On the contrary, to obtain ultrashort pulses with light it is necessary to develop sources with broad emission spectra. The study of the conditions under which various spectral distributions are produced in SRS of light is therefore of practical interest.

As a result of the rapid development of nonlinear optics, the main properties of SRS of light (the distribution of the energy among the components and in space, the angular directivity of the radiation) have been investigated in sufficient detail in the last few years. The least investigated is the spectral distribution of the intensity of the SRS components. Experiments performed by the time the present research was initiated have followed two practically independent trends: a) investigation of the broadening of the spectral distribution of the intensity of the first SRS Stokes component and principally of the exciting laser radiation in the region of self-focusing of light in media having large Kerr constants; b) measurements of the spectral width of the first SRS Stokes component and comparison with the theoretical expression that takes into account the narrowing of the spectral line width of the Raman scattering when the scattered light passes through a nonlinear medium. There

---

*Dissertation for the degree of Candidate of Physicomathematical Sciences, defended November 28, 1975 at the Session of the Scientific Council of the Department of General and Applied Physics, Moscow Physicotechnical Institute. The advisers were Doctor of Physicomathematical Sciences M. M. Sushchinskii and Candidate of Physicomathematical Sciences A. I. Sokolovskaya.

are also some published papers that cannot be identified with either trend. Their authors observed in the spectral distribution of the intensity of the first SRS Stokes component a complex structure consisting of individual components, the distances between which ranged from fractions of a reciprocal centimeter to several reciprocal centimeters. The origin of this spectral structure is not fully clear. It should be noted that the interpretation of the experimental results was made difficult by the fact that in most studies the excitation source was a multimode laser.

Spectral broadenings of the exciting radiation and of the first SRS Stokes component on the order of 10-100 cm^{-1} was observed in a number of substances with large Kerr constants, excited by a laser with one or several longitudinal modes [1-11]. The cause of the large spectral broadenings was theoretically attributed to self-modulation of the laser beam and of the SRS in the self-focusing filaments produced in the laser beam [1-3, 8] or to beats of the modes of the multimode laser [9-11]. The question of the spectral broadening of laser radiation and of the SRS components was considered in greatest detail by Lugovoi and Prokhorov [12] on the basis of the hypothesis that the light propagates in the medium in the form of moving focal regions. It should be noted that the theoretical understanding of the question of the onset of large spectral broadenings of laser emission and of SRS progressed much farther than the experiment. The experimental data, on the other hand, were quite incomplete. The reported studies noted principally the very fact that spectral broadening of the exciting radiation takes place, and sometimes broadening of the first SRS Stokes component at the points of self-focusing inside the medium in a plane near the exit end of the cell, while other papers reported that the spectral composition of the light leaving the medium becomes more complicated. The published papers did not contain sufficient information that would make it possible to establish which of the proposed physical mechanisms leads to the broadening of the spectral distribution of the radiation. Data on simultaneous observation of the spectral distribution of the intensity of the SRS components and of the exciting radiation were missing from most papers. In a number of cases only the spectral distribution of the intensity of the light leaving the medium was registered. No experimental observations of self-focusing were made in that case, so that the role of the latter remained unknown. In those cases when the spectral distribution was observed inside the medium, in a plane near the exit end of the cell, the excitation conditions corresponded to the regime of saturation of the SRS intensity. It was then impossible to determine whether the large spectral broadenings of the SRS components were produced at a definite threshold pump intensity or not. The experiments did not clarify the conditions under which it is possible to observe in the near field of the higher-order SRS components radiation cones, or self-focusing without spectral broadening and with large spectral broadenings.

In reports of the measurements of the width of the spectral distribution of the first SRS Stokes component [13-18], various authors cite rather contradictory data. When the experimental results are compared with a theory that takes into account the narrowing of the Raman-scattering line on passing through a nonlinear medium, many authors find that the experimental data deviate from the theoretical estimates. According to the data of other authors, agreement between theory and experiment is observed. No quantitative estimates of the line width of the SRS components based on the theoretical formulas were given. The reason was the absence of experimental measurements of the SRS gain, and in a number of studies the absence of experimental measurements of the spectral width of the Raman-scattering lines. It should be noted that the spectral distribution of the higher SRS Stokes and anti-Stokes components were hardly investigated. In this connection, the question whether it is legitimate to describe the SRS line widths by means of theoretical formulas obtained under the assumption that the Raman scattering lines become narrower when the scattered light passes through a nonlinear medium remained unanswered.

The purpose of the present study was a systematic investigation of the contour and width of the spectral distribution of the intensity of SRS components of media with substantially differing Kerr constants under various excitation conditions, with an aim of obtaining a complete picture of the conditions under which various spectral distributions of the intensities occur, and to reconcile the contradictory results of earlier studies. A simultaneously undertaken task was to obtain information on the influence exerted on the spectral distribution of the intensity of the SRS components by other nonlinear processes (primarily self-focusing of light) which are produced in media on passage of powerful laser radiation.

CHAPTER I

REVIEW OF EXPERIMENTAL AND THEORETICAL INVESTIGATIONS
OF THE SPECTRAL DISTRIBUTION OF THE INTENSITY OF THE
COMPONENTS OF SRS OF LIGHT

In the survey of the literature we shall dwell on the principal experimental and theoretical papers that describe the opinions held at the time this work was started concerning the spectral distribution of the intensities of the SRS components.

The first theoretical premises concerning the spectral distribution of the intensity of the components of SRS of light were given in papers by Lugovoi [19], Sushchinskii [20], Bloembergen [21], and D'yakov [22]. It was assumed that the "bare" source of the SRS is spontaneous noise — ordinary Raman scattering (RS) of light. The intensity of the SRS increases exponentially as a function of the noise in accordance with the equation [19, 23, 24]

$$I_s(l) = I_s(0) \exp(glI_0),$$

where $I_S(l)$ is the SRS intensity; $I_S(0)$ is the RS intensity; I_0 is the intensity of the incident laser radiation; l is the total length of the medium; g is the gain defined in this case by the formula [25]

$$g = \frac{2c^2}{\pi h n^2} \frac{N}{\gamma_{RS}(\nu_0 - \nu_R)^3} \frac{d\sigma_{\parallel}}{d\Omega} \tag{1}$$

(amplification of radiation polarized in the same plane as the laser radiation). Here c is the speed of light; h is Planck's constant; n is the refractive index; N is the effective number of molecules in cm^3; γ_{RS} is the width of the RS line; ν_0 is the frequency of the exciting radiation; ν_R is the frequency of the molecular oscillation; $(\nu_0 - \nu_R)$ is the frequency of the RS line; $d\sigma_{\parallel}/d\Omega$ is the total differential cross section of the RS per molecule, per steradian, and for one polarization.

On passing through a nonlinear medium, the RS line contour becomes deformed. The line of the scattered radiation then becomes narrower (compared with the RS), and the line width of the SRS in the region of the exponential growth of the intensity of the first SRS Stokes component is determined by the formula (see, e.g., [19, 20])

$$\gamma_{SRS} = \gamma_{RS}(gl)^{-1/2}, \tag{2}$$

where γ_{SRS} is the width of the SRS spectral line; γ_{RS} is the width of the RS spectral line.

Beyond the region of exponential growth of the intensity of the SRS components, saturation sets in. If the saturation is reached in one pass through the active medium, then we can estimate the width of the line of the m-th SRS component when it saturates. D'yakov [22] obtained the following estimate:

$$\gamma_{SRS}^{(m)} = \gamma_{RS} \sqrt{\frac{m \ln 2}{\ln(1/a)}} ; \qquad I_{thr}^{(m)} = \ln(1/a) \cdot (g_m l_j)^{-1}, \tag{3}$$

where I_{thr} is the threshold value of the pump intensity for excitation of the m-th SRS component; g_m is the gain of the corresponding component; a is the initial relative level of the scattering intensity [26].

The number of experimental papers devoted to a check on the validity of (2) is small [13-18]. Under the neatest experimental conditions, measurement of the width of the first SRS Stokes component was carried out by Chirkov and co-workers [13]. The SRS was excited by a laser that generated a single longitudinal mode. The line widths of the first SRS Stokes component of stilbene, calcite, potassium nitrate, benzene, and acetone were measured. The experimentally observed line widths were compared with theoretical expressions. The SRS line width was either on the order of the theoretical value or somewhat smaller.

Unfortunately, the theoretical estimates of γ_{SRS} were only approximate, because of the lack of data on the SRS gain, and in many cases due to lack of data on the RS line widths for the indicated substances. It was proposed in [13] that $gl \sim 20-40$ for the investigated substances. Then the width of the SRS, according to (2), should be smaller than the line width of the corresponding RS oscillation by a factor 5-7. It was also noted in [13] that when the temperature is lowered the line width of the first SRS Stokes component of stilbene powder and potassium nitrate powder decreases.

The dependence of the line width of the first SRS Stokes component of stilbene powder (1593 cm^{-1}) on the energy of the exciting radiation was investigated in [18]. It was observed that from the threshold in the region of the fast growth of the intensity of the first SRS component there is observed a negligible increase of the line width (from 0.33 to 0.4 cm^{-1}), and subsequently, in the saturation region, the width remains practically unchanged. A substantial difference was noted in [18] between the experimentally observed width of the first SRS component of stilbene and the value obtained from formula (2). This conclusion, however, has no real basis, since no quantitative comparison of the obtained experimental data with formula (2) were made, owing to the lack of data on the gain in stilbene, and the changes in the line width were quite small.

In [14-17] they also measured the line widths of the first SRS components of a number of substances, and a decrease of the SRS line width compared with the line width of the corresponding ordinary RS was observed.

The question of the onset of large spectral broadenings of the components of the SRS of light was considered independently in the literature.

It was noted in a group of papers by Bloembergen and Lallemand [9-11] that a considerable broadening takes place of the Stokes and anti-Stokes SRS components of liquids with anisotropic molecules if the exciting laser beam is not **strictly monochromatic.** They used an exciting-radiation source in the form of a ruby laser that generated two modes of space 1.6 cm^{-1} apart. The spectral distribution of the intensity of the first Stokes component of the SRS of benzaldehyde contained more than 50 bands. The dependence of this spectral distribution of the intensity of the SRS component on the temperature was investigated. It turned out that the largest broadening of the spectrum of the first SRS Stokes component is observed at high temperatures, when the threshold power for self-focusing is maximal, and the coefficient of conversion of the exciting radiation into SRS is the lowest. The width of the spectral distribution at +85°C is five times larger than at −25°C. The spectral broadening was maximal for molecules with maximum anisotropy (carbon disulfide), and the broadening decreased when the relaxation time of the molecule increased. The spectral distribution of the radiation emerging from the medium was observed in these studies. No experimental observations were made of the self-focusing of the radiation in the investigated substances, nor was there a study made of the spectral composition of the laser radiation passing through the medium.

In a number of papers on the influence of self-focusing of light on the spectral distribution of radiation passing through a nonlinear medium, mention was made of large broadenings of the first SRS Stokes component occurring at the self-focusing dots [1, 2], both upon excitation by a multimode laser and by a single-mode laser. Principal attention was paid in these studies to the spectral distribution of the laser radiation at the self-focusing dots [1-8]. All the theoretical papers in which various physical mechanisms were proposed dealt with rare exceptions with the exciting radiation. The role of the scatterings (SRS, SMBS) remained unclear.

We now dwell briefly on the main publications that describe the presently prevailing theoretical premises concerning the origin of the spectral broadening of radiation passing through a nonlinear medium. The appearance of spectral broadenings of laser radiation (and sometimes Stokes radiation) is explained theoretically by most workers as being due to phase modulation of the laser light in the light-conducting filaments produced as a result of self-focusing of the laser radiation in the medium [1-8].

It is known that when laser radiation passes through a liquid dielectric, the dielectric constant changes, and consequently also the refractive index, and this leads to a deflection of the light towards the larger value of n. If this refraction cancels the diffraction divergence, a waveguide is produced, in which the laser radiation propagates with a high power density. In liquids the principal mechanism that leads to self-focusing is the Kerr effect. The development of a theory of self-focusing is the subject of a large number of papers [27-32]. The concept of critical beam power (P_{cr}) at which self-focusing of light sets in was introduced. The universally accepted point of view was reduced to the notion that the point of collapse is followed by **self-trapping** of the beam into waveguide propagation. Under these conditions, the expressions obtained for the amplitude and phase of the laser wave propagating in the medium are similar to those given below [12]:

$$|\vec{E}(0, z, t)| \simeq \left| E_0\left(t - \frac{z}{v}\right)\right|,$$ (4)

$$\overline{\varphi}(0, z, t) \simeq \frac{1}{2}n_2 kz E_0^2\left(t - \frac{z}{v}\right) - \frac{z}{k\bar{a}_0^2},$$ (5)

where $\bar{a}_0$ is the radius of the distribution of the intensity in the paraxial zone; v is the speed of light in the medium; n_2 is the nonlinear part of the refractive index; k is the wave vector; z is the coordinate along the beam axis.

It is seen that with respect to the envelope (4) the propagation of the pulse is similar to that in a linear medium. On the other hand, the phase $\overline{\varphi}$ at $n_2 \neq 0$ remains dependent on the time and is modulated in accordance with the time variation of the power in the initial cross section of the incident beam. The depth of modulation is proportional to the thickness of the layer of nonlinear medium traversed by the pulse. The phase modulation leads to a change in the spectrum of the pulse propagating in the medium [1-11]. Shimizu [3] expresses the total width of the obtained spectral distribution in the form

$$\Delta\omega \simeq \frac{1.3 \; kz n_2 E_{0\,max}^2}{\tau_p},$$ (6)

where $E_{0\,max}^2$ is the maximum amplitude E_0 during the time of the pulse; τ_p is the pulse duration.

If it is recognized that the width of the spectral distribution of the intensity of the incident beam is $\sim 1/\tau_p$, then Eq. (6) yields directly the broadening $\Delta\omega\tau_p$ of the spectrum of the pulse propagating in the nonlinear medium. In the case of phase modulation, the broadening of the spectrum is proportional to the path z of the pulse in the medium [24]. It is seen that the broadening is proportional to the quantity $n_2 E_0^2{}_{max}/\tau_p$, i.e., to the rate of change of the refractive index with time (the so-called pseudo-Doppler effect). The possibility of Doppler broadening of the spectrum of the pulse on passing through a nonlinear medium was indicated by Zel'-dovich and Raizer in [31]. Shimizu [3] has shown that such a process can also lead to broadening of the pulse spectrum, and that the same value of the instantaneous frequency $\omega_1(t)$ is realized for two instants of time $t_1 \neq t_2$, and at the phase difference $(2m + 1)\pi$ (m is an integer) the distribution of $I(\omega)$ has near ω_1 a minimum, while at a phase difference $2m\pi$ it has a maximum. Consequently, the frequency distribution of the intensity $I(\omega)$ has an oscillatory form with a large number of maxima and minima [1, 3, 8]. The period of these oscillations can be estimated at [12]:

$$\Delta\Omega = 2\pi\Delta\omega/|\Delta\overline{\varphi}| \simeq 8/\tau_p.$$

If the broadening of the spectrum is large enough, an important role is assumed by the dispersion of the linear part of the refractive index, as a result of which the phase relations are disturbed and the phase modulation can turn into amplitude modulation. Further broadening of the spectrum can result from beats between different frequency components. An analogous picture was drawn in the papers of Bloembergen and co-authors [9-11] for the mechanism of broadening of laser spectra in the case of beats between different frequency components of a multimode laser. Most authors note that experiment reveals weaker anti-Stokes sidebands [1-3, 8-11], i.e., the broadening of the spectrum is asymmetric. The asymmetry of the spectral distribution of the intensity cannot be adequately explained if it is assumed that in the case of self-focusing the laser radiation produces in the medium light-conducting "filaments." The authors of the cited papers invoke various additional assumptions to explain the observed asymmetry. Thus, for example, it is assumed in [2] that the asymmetry appears in the spectrum as a result of distortion of the waveform of the laser pulse in the medium (the trailing edge becomes steeper and the leading edge more gently sloping). The energy in the spectrum then goes over into the Stokes region. Bloembergen and co-authors [9-11] believe that the asymmetry is the consequence of stimulated scattering in the Rayleigh line wing [33] (gain in the Stokes region of the spectrum and loss in the anti-Stokes region). These assumptions, however, do not present a complete picture of the observed asymmetry in spectra, and in some cases are even incapable of explaining the observed picture (for example, when the greater part of the broadening lies in the anti-Stokes region [34, 35]).

Dyshko, Lugogoi, and Prokhorov [36], on the basis of a numerical solution of the problem of the propagation of a light beam with Gaussian intensity distribution in a space with a planar initial phase front in media with Kerr nonlinearity, have proposed a new picture of the development of self-focusing. This theory explains many hitherto ununderstandable results of experiments (the appearance of ultrashort SRS pulses, the asymmetry of the broadenings, the angular directivity of the higher-order components).

According to [36], the collapse point is not the start of the waveguide filament, but the center of the first focal region. Not the entire power of the beam flows into this focal region, but only a value close to critical (P_{cr}). Part of the energy is absorbed in the focus because of nonlinear absorption, and the remainder emerges in the form of a rapidly diverging ring-shaped wave. The part of the beam that moves on the sides past the first focus produces in exactly the same manner, at a certain distance from it, the next focus. The power flowing into this focus is also P_{cr}. Similarly, part of the energy is absorbed in the focus, and part emerges in the form of a ring-shaped wave. The remaining foci are formed analogously. The formation of the foci continues until the entire power of the initial beam is consumed. The focus-formation process then stops, and therefore the total number of foci depends on the initial power of the beam.

Under real conditions, the incident beam is not stationary. The power of this beam varies with time in accordance with the envelope of the laser pulse. Since the positions of the foci on the beam axis depend on the initial power, and this power itself varies with time, this leads to motion of the foci along the beam axis [37]. These concepts were confirmed by the experiments of Korobkin and Serov [38]. They have shown that the experimentally observed waveguides are the integral picture of focal regions that travel at high velocity. For giant laser pulses and under typical conditions, the velocities of the foci in the medium have a characteristic singularity: the m-th focus initially appears in the section $z = l$ and moves into the interior of the considered layer. The focus then stops at the point $z = \xi_{min}^{(m)}$ and moves subsequently towards the boundary $z = l$. Naturally, the focus stays the longest at the turning point $z = \xi_{min}^{(m)}$. One can therefore expect the processes occurring in the medium at high power densities, but requiring a sufficient development time, to manifest themselves primarily at or near the turning points of the foci [12, 37].

If the initial beam power greatly exceeds the critical value (by several orders of magnitude), the deviations from axial symmetry of the initial distribution in the real beam can in themselves contain a supercritical power and by the same token autonomous multifocus structures can be produced in individual regions of the beam. This picture of the breakup of the beam at large supercritical powers was observed in experiment [38]. The theoretical possibility of this breakup of the beam was indicated in [30, 35, 39].

Thus, a consistent theory of the propagation of high-power laser beams in media with Kerr nonlinearity leads to a multifocus picture of this propagation and to the onset of moving foci. Lugovoi and Prokhorov [12] calculated the emission spectrum due to the motion of one of the focal regions of a multifocus structure [12, 40], while Wang and Shen [41] verified this calculation experimentally for carbon disulfide.

It should be noted that the interference of the radiation from two or more focal regions can obviously lead to a quasiperiodic structure in the summary frequency distribution of the intensity. Both a continuous and a quasiperiodic spectrum is observed in experiment, depending on the excitation conditions.

When the focal region moves toward the turning point, its dimensions can either be increased by the SRS, or the region can vanish at a certain instant of time. This takes place when the absorption due to the SRS becomes equal to the inertialess (for example, three-photon) absorption. An ultrashort SRS pulse is excited at that instant [42]. The duration of the corresponding pulse of the first SRS Stokes component is on the order of the length of the focal region (l_{fp}) divided by the speed of light in the medium, i.e., $l_{fp}/v \approx 10^{-11}$ sec. The ultrashort SRS pulse excited in this manner will propagate in the medium in the opposite direction (backwards) to the velocity of light in the medium. Ultrashort SRS pulses in the backward direction were first observed experimentally by Maier, Kaiser, and Giordmaine [43], and later by Loy and Shen [44].

If the effective transformation of the energy of the focal region into SRS takes place near the turning point or at the turning point itself, i.e., $v_{fp} \ll v$ (v_{fp} is the velocity of the focal region), then the pulses of the first SRS Stokes component should be produced to an equal degree in the forward and in the backward directions. Such ultrashort pulses were observed experimentally in [42], as well as in [45]. The width of the spectrum $\Delta\omega$ at these durations [12] can be estimated. Thus, the result for $\tau_p = 3 \cdot 10^{-11}$ sec is $\Delta\omega = 200$ cm^{-1}.

As already stated above, the experimental and theoretical papers dealt mainly with the development of self-focusing of laser radiation passing through a medium. SRS was considered in a number of papers as a phenomenon that leads to losses and to destruction of the regions of self-focusing of the laser radiation. Recently Sokolovskaya and co-authors [17, 46-49] have shown that in substances with large nonlinear Raman susceptibilities, the SRS of the light undergoes self-focusing and produces a multifocus structure independently of the exciting radiation. These results pertained to the first SRS Stokes component, and the higher-order components were not investigated. Nor was the connection between the self-focusing of the SRS and the spectral distribution of the intensities of the components investigated.

There are published reports of observation of the SRS-component spectral intensity distributions not similar to those considered above. These include primarily studies [50, 51] in which a structure consisting of individual components (from 1-2 to 5-6) was observed in the first SRS Stokes component of carbon disulfide, styrene, isoprene, benzene, 1,3-pentadiene, and nitrobenzene near the threshold. The spacing between components was 10-12 cm^{-1}. The SRS was excited by a multimode laser, and the feedback was not eliminated from the optical system. The Q-switching of the resonator was by a rotating prism, so that the SRS could be excited by several spikes of the ruby laser. The observed spectral picture was attributed in [50, 51] to the longitudinal Doppler effect.

Brewer [4] observed, at the self-focusing dots, components shifted towards the red end of the spectrum in the spectral distribution of the intensity of the exciting radiation passing through carbon disulfide or nitrobenzene excited by a parallel beam. The shifts varied by an approximate factor of five from one self-focusing dot to another. The average shift for carbon disulfide was approximately 11 cm^{-1}. The components lie in the vicinity of the frequency of the stimulated Rayleigh-wing scattering [33], and the shift for carbon disulfide is on the order of 3 cm^{-1}. In nitrobenzene, smaller shifts were observed, on the order of 0.032, 0.057, 0.093, and 0.125 cm^{-1}. The observed frequency shifts do not agree with the intermode spacing in the laser (0.01 cm^{-1}). The shifts in the nitrobenzene are much less than in carbon disulfide, this being due, as proposed by Brewer, to the smaller change of the refractive index in nitrobenzene and to the difference between the relaxation times ($\tau_{C_6H_5NO_2} > \tau_{CS_2}$ by approximately 30 times). Brewer believes that the frequency components observed by him correspond to stimulated two-photon and four-photon Rayleigh-wing scattering as a result of the molecular orientational Kerr effect. He believes this scattering to be caused by self-focusing.

A complicated structure in the spectrum of the SRS components of aqueous solution of HNO_3 ($\nu_R = 1324$ and 1060 cm^{-1}) was observed in [52]. The authors believe the cause of the spectral splitting of the SRS components to be the anharmonicity of the molecular vibrations, which is the result of the loss of thermodynamic equilibrium in the medium caused by the powerful laser pulses [53].

A structure in the spectrum of the first SRS component of calcite was observed in [54]. The widths of the individual components were on the order of the line width of the exciting radiation (0.02 cm^{-1}). The origin of this structure and its singularities remained unexplained.

An analysis of the published papers shows that at the time the present study was started the experimental material was quite incomplete and did not provide a clear picture of the spectral distribution of the intensity of the SRS components in media in the condensed state.

It was therefore proposed to investigate the spectral composition of the SRS components in a number of substances with different optical characteristics, using an instrument with high spectral resolution and using excitation by a laser that generates one longitudinal anode. It was of interest to clarify the question of the minimal spectral width of the SRS line.

To check on the validity of the theoretical expressions (2) and (3) for the description of the SRS line width we investigated the contours and widths of the SRS components at various excitation intensities. The program included simultaneous measurements of the width of the first SRS Stokes component in the direction of the exciting radiation and in the opposite direction.

The influence of the temperature of the scattering medium on the contour and width of the spectral distribution of the intensity of the SRS component was also investigated.

In addition, it was proposed to systematically investigate the conditions under which the SRS components undergo large spectral broadenings, and their connection with the development of self-focusing of light in the medium. This called for investigations of substances having both large (carbon disulfide, nitrobenzene) and small (calcite, liquid nitrogen) Kerr constants, and simultaneous observation of the development of self-focusing in substances and of the spectral and spatial distribution of the intensity of the exciting radiation and of the SRS at various excitation energies.

CHAPTER II

SETUP AND PROCEDURE FOR THE MEASUREMENT OF THE SPATIAL
AND SPECTRAL CHARACTERISTICS OF SRS

1. Source of Exciting Radiation and Setup for the Experimental
Investigation of the Spectral Distribution of the SRS
Component Intensity

The SRS of light was excited with a giant pulse from a ruby laser. The Q-switch was a passive shutter in the form of a solution of cryptocyanine in ethyl alcohol (1,1'-diethyl-4,4'-dicarbocyanine iodide). Our experience has shown that the mode composition of the radiation depends strongly on the quality of the ruby crystal. We therefore selected, from a large batch, two or three ruby rods with the best optical homogeneity. These crystals were subjected to diffusion annealing for 40 h to relieve the residual stresses. When these ruby rods were used to obtain a single longitudinal mode in the laser emission, the selective action of the passive shutter and of the resonator mirrors was sufficient. The ruby crystal was usually 120 mm long, 7-8 mm in diameter, and had sapphire end pieces. The parameters of the emission of the ruby laser employed by us were the following: divergence 3.5', pulse duration at half height 20 nsec, lasing line width 0.015 cm^{-1}, maximum power in the pulse 10 MW. Several experiments required a higher exciting-radiation power. In this case the laser pulse was amplified with a one-stage amplifier. The active element in the amplifier was a ruby crystal 240 mm long, 16 mm in diameter, and with sapphire end pieces. The end faces of the crystal were polished and inclined at the Brewster angle.

Our investigations were carried out with a ruby laser with specially stabilized emission. To this end the ruby crystals of the amplifier and of the laser were maintained at constant temperature by liquid-nitrogen vapor flowing through the illuminators. The rate of nitrogen evaporation could be regulated. The temperature in the illuminator was monitored with a copper−constantan thermocouple connected to an M-95 microammeter.

The stability of the spectral composition and of the duration of the ruby pulse was monitored in each flash with a Fabry−Perot etalon having a base 150 mm and with an FÉK-15 photocell connected to an I2-7 oscilloscope. The number of spike in the laser pulse was monitored by a photodiode connected to an S1-18 oscilloscope. The laser emission energy was measured with a vacuum thermocouple and an F116/1 galvanometer whose scale was calibrated with an IMO-2 calorimeter.

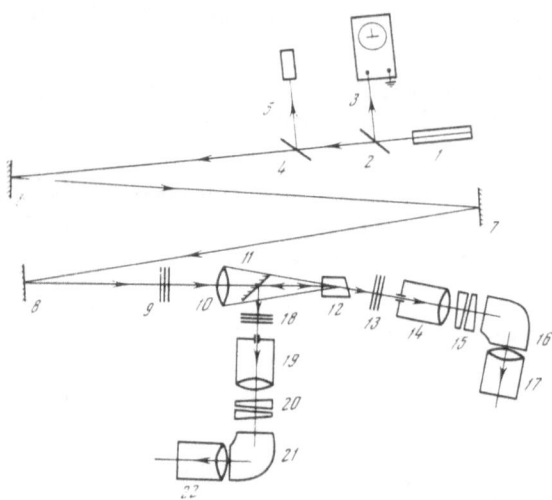

Fig. 1. Optical diagram of the setup for the study of the spectral distribution of the intensity of the SRS components. (1) Laser; (2, 4) turning plates; (3) photodiode connected to oscilloscope S1-18; (5) thermocouple; (6-8) 100% mirrors of the optical delay system; (9, 13, 18) stacks of light filters; (10) lens with f = 420 mm; (11) 30% mirror; (12) sample; (14, 19) collimators of ISP-51 spectrograph; (15, 20) Fabry−Perot etalons; (16, 21) prism of the ISP-51 spectrographs; (17, 22) cameras of spectrographs.

The objects of the investigation were substances with substantially differing Kerr constants. Table 1 lists the wavelengths of the lines observed by us in experiment in the SRS spectra of these substances.

We investigated the spectral composition of the components of SRS of light propagating in the direction of the exciting radiation ("forward") and in the opposite direction ("backward"). The experimental setup is shown in Fig. 1. The emission of laser 1 was directed to the sample 12. Turning plates 2 and 4 directed part of the exciting radiation, to monitor the excitation energy and the number of spikes in the lasing pulse, to photodiode 3, connected to an oscilloscope, and to thermocouple 5. To eliminate feedback between the laser and the sample, the optical path was lengthened with the aid of mirror system 6-8, so that the delay time between the lasing pulse and the pulse reflected from the end face of the cell amounted to 40 nsec. The exciting radiation was then focused with a long-focus lens 10 approximately on the center of the investigated sample. To eliminate the influence of the feedback inside the sample itself, resulting from reflections from the end faces, the crystal end faces (or the cell windows) were misaligned by an angle 10-12°. The samples with parallel end faces were placed in the beam of the exciting radiation at an angle. The energy of the exciting radiation was varied with a stack of calibrated neutral light filters 9. The recording system consisted of an ISP-51 spectrograph crossed with a Fabry−Perot etalon, which was placed between the collimator and the prisms in positions 15 and 20.

The radiation scattered in the forward direction was incident on the slit of the ISP-51 spectrograph (14, 16, 17). The radiation scattered in the backward direction was directed by a turning mirror with reflection coefficient 30% (11) to the slit of another spectrograph ISP-51 (19, 21, 22), likewise crossed with Fabry−Perot etalon 20. The width of the spectrograph slit was 2 mm. The center of the interference pattern was located in the center of the image of the slit. For different SRS frequencies, the center of the interference pattern remained unchanged in position relative to the image of the slit − this is the principal advantage of such a placement of the etalon in the spectrograph.

When the investigation of the spectral composition of the SRS called for observation of the spectral distribution of the intensity of the components in a wide wavelength range, we used an STE-1 spectrograph whose linear dispersion in the investigated region ranged from 5.6 to 12.8 Å/mm. To obtain normal densities on the photographic plate, light-filter stacks 13 and 18 were placed in front of the spectrograph slit. Depending on the investigated region of the spectrum, we used either soviet photographic plates (infra 740, 780, 840) or photographic plates made by "ORWO" (I-750, I-850, I-950, I-1050, RP-1, WP-1, WP-3).

TABLE 1

Substance	$\nu_R \cdot cm^{-1}$	SRS component wavelength, Å					
		3aSt	2aSt	1aSt	1St	2St	3St
N_2	2329.7	4670	5246	5967	8283	10264	—
$CaCO_3$	1085.6	5663	6034	6456	7509	8172	8970
CS_2	656	6107	6364	6641	7274	7639	8405
$C_6H_5NO_2$	1345	—	5847	6350	7658	8537	—

The calcite single crystal was oriented such that its optical axis was parallel to the direction of propagation of the exciting radiation. A dewar with windows was used to investigate liquid nitrogen.

To investigate the spectral composition of the SRS components at different temperatures of the investigated substance, we used cells of special design. To work at temperatures from 300 to 500°K, the sample was placed in a ceramic jacket heated by a copper coil regulated by an autotransformer. The temperature gradient between the center and the edges of the sample did not exceed 0.5°.

The scattering medium was cooled to below room temperature in the following manner. The sample was placed in a special jacket covered by a thermal-insulating layer through which liquid-nitrogen vapor was passed. The sample temperature was measured with a calibrated copper—constantan thermocouple. The calcite crystal was cooled to 77°K by placing the crystal in contact with the liquid nitrogen in a dewar with plane-parallel windows. To prevent SRS from occurring in the liquid nitrogen, only the lower face of the crystal was completely immersed in the nitrogen.

2. Procedure of Measuring the Contour and Spectral Width

of the SRS Component Distribution

The procedure for the measurement of the contours and widths of the lines of the molecular spectra was developed in the G. S. Landsberg Optical Laboratory of the Physics Institute of the Academy of Sciences and was described in detail, for example, in [55]. It was this procedure which was used by us in the investigation of the contours and widths of the SRS lines. The SRS spectra on the photographic plates were recorded with the two-beam recording microphotometer IFO-451. The photographic densities were converted into intensities with the aid of density markers. To obtain the density markers at the required wavelength, an attenuator, consisting of seven steps with different transmission, was placed on the spectrograph slit, illuminated with the light of the SRS, and photographed on a photographic plate. To obtain uniform illumination in the plane of the slit, a ground-glass plate and a scattering lens were placed in front of the attenuator. The uniformity of the illumination of the slit was verified with a microphotometer. In a number of cases, a Zeiss comparator was used to estimate the width of the spectral distribution and to determine the distances between the components.

The SRS-component spectral line width was determined by the method of squares when the spectrum was obtained with a Fabry—Perot etalon [56, 57].

The condition that determines the maximum in the interference pattern for the Fabry—Perot etalon for a given frequency ν is of the form

$$2\nu\mu t \cos i = m, \tag{7}$$

where μ is the refractive index of the medium between the plates of the etalon; t is the distance between the plates, m is the number of the interference order, f is the focal length of the lens in whose focal plane the interference pattern is viewed, and i is the angle at which the ring is obtained in the focal plane of the lens.

The angle i is small, therefore i = D/2f, where D is the diameter of the ring on the interferogram.

We differentiate expression (7)

$$\frac{di}{d\nu} = \frac{1}{\nu \operatorname{tg} i} = \frac{1}{\nu i},$$

$$i\,di = d\nu/\nu. \tag{8}$$

Let $\Delta\tilde{\nu} = \tilde{\nu}_1 - \tilde{\nu}_2$ and ν be a certain average value of ν from the interval $\Delta\nu$; ν_1 and ν_2 produce rings on the interferogram within the limits of one order. Then, integrating (8) over the interval di, we obtain

$$\frac{i_1^2 - i_2^2}{2} = \frac{\Delta\nu}{\tilde{\nu}}.$$

Substituting in this expression the angles represented in terms of the diameter of the ring and the focal length of the lens, we obtain

$$\Delta\nu = \tilde{\nu}\,\frac{D_{m\nu_1}^2 - D_{m\nu_2}^2}{8f^2}. \tag{9}$$

As seen from (9), to determine the width of the line (or the distance between the components of the fine structure of the line) it is necessary to measure the corresponding diameters of the rings and the interferogram.

When the line contours were smooth in form, the width was determined by measuring the diameters on which were located the points at the half-intensity level of the interference ring.

3. Determination and Account of the Instrumental Function

It is known that real spectral instruments are not ideal harmonic analyzers of the radiation, i.e., the distribution of the energy over the spectrum, obtained with the aid of the spectral instrument, differs from the true distribution given by the Fourier expansion of the investigated radiation. Some of the distortions introduced by the spectral instrument are slowly varying and large in scale. Examples are the parasitic background illumination produced on the spectrogram by scattering from the elements of the instrument, which adds some slowly varying amount to the true distribution of the intensity, variation of the transmission of the instrument with wavelength, the dispersion of the instrument, the dispersion of the angular and linear magnification, the sensitivity of the receiver of the radiation, and others. These distortions can be easily taken into account since methods have been developed for this purpose [55, 56].

An important role is played in the investigation of the fine structure of the line by distortions of another type. These distortions are due to the fact that even in monochromatic radiation a real spectral instrument gives a certain spectral energy distribution with a finite width. The form of this distribution and its width are determined by various causes: diffraction by the diaphragms of the optical system of the spectral instrument, aberrations of the system, the finite width of the slits, scattering in the sensitive layer of the photographic plate, and others. The distribution width obtained in a spectral instrument under monochromatic illumination is called the instrumental width (function) of the instrument. The question of the instrumental function was developed in detail by Rautian [58]. The existence of the instrumental function determines for the most part the capability of the instrument in investigations of fine details or the structure of the spectrum. When measuring the line widths and investigating the structure of the contour it is therefore necessary to know the instrumental width, so as to obtain real estimates of the measured quantities and of the capability of the apparatus.

The instrumental function of a spectrograph is determined to a considerable degree by the scattering in the photographic emulsion [59]. The instrumental widths of the photographic emulsions which we used was on the order of 0.015 mm (which for our installation is less than the instrumental width of the Fabry–Perot etalon, so that in our case the instrumental width was determined by the width of the instrumental function of the Fabry–Perot etalon). The instrumental function of the Fabry–Perot etalon is the well-known Airy function [56, 57]. We have estimated the instrumental width for our installation experimentally. It is known that for a given Fabry–Perot etalon the following relation is valid [57]:

$$a/\Delta F = \text{const}, \tag{10}$$

where a is the instrumental width and ΔF is the dispersion region of the etalon.

If the width of the ring in the interference pattern is smaller by at least a factor of five than the distance between the orders, then this width is determined mainly by the instrumental width a [57]. If we obtain the interferogram of a ruby laser with a line width 0.015 cm^{-1} through an etalon with a base 0.3 mm (distance between orders 16.7 cm^{-1}), then the condition cited above will be satisfied. Thus, we can assume with high accuracy that in this case the width of the interference ring is determined only by the instrumental width and we can determine the value of a for a given pair of mirrors. It is possible next to estimate the instrumental width of this pair of mirrors for any value of ΔF, using relation (10). The instrumental value was determined by us in the same experimental geometry as the remaining measurements, and amounted to less than 10% of the measured spectral width.

4. Procedure for Observation of Self-Focusing and of the Spatial Distribution of the Intensity of the SRS Components

To investigate the self-focusing of light in substances through which intense laser radiation passes, we have observed the distribution of the intensity over the cross sections of the laser beam and of the SRS components inside the medium near the entrance and exit windows of the cell (or end faces of the crystal). In a number of cases we investigated the intensity distribution in a plane in the interior of the cell. The experimental setup is shown in Fig. 2. The laser light was focused by lens of focal length f = 420 mm into the interior of the medium. The plane 4 inside the medium was projected by a lens of focal length f = 100 mm, with fivefold magnification, on the slit of an STÉ-1 spectrograph, after which the slit was removed. We obtained in the plane of the photographic plate the distribution of the intensity in the plane 4 for the exciting radiation and for the SRS

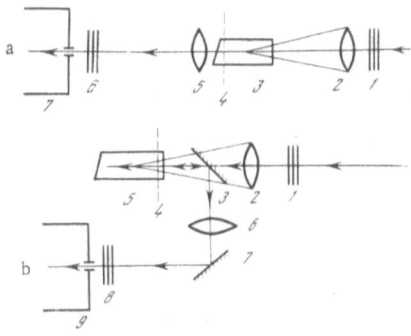

Fig. 2. Optical system of the setup for the investigation of
the distribution of the intensity in the cross sections of the
beams of the exciting radiation and of SRS components inside
the medium near the cell windows. In the forward direction
(a): (1) stack of neutral filters; (2) lens with f = 420 mm; (3)
sample; (4) plane projected on the slit; (5) lens with f = 100
mm; (6) stack of light filters; (7) STÉ-1 spectrograph. In the
backward direction (b): (1) light filters; (2) lens with f = 420
mm; (3) 30% mirror; (4) plane projected on the slit; (5) sample; (6) lens with f = 100 mm; (7) 100% mirror; (8) light filters; (9) STÉ-1 spectrograph.

components. Figure 2a shows the setup for the investigation of light scattering in the forward direction, and
Fig. 2b shows the same in the backward direction. The energy of the exciting radiation was varied with a stack of
calibrated neutral light filters 1. The resolution of the system was determined experimentally with the aid of
a set of targets and amounted to 10 μm, while the depth of focus was less than 1 mm.

To investigate the angular distribution of the intensity of the SRS components we used a setup in which
the scattered radiation emerging from the sample was focused by a lens of focal length f = 100 mm on the slit
of the spectrograph in such a way that the focal plane of the lens coincided with the plane of the spectrograph
slit, after which the slit was removed. In the plane of the slit, and hence in the plane of the photographic plate,
we obtained the angular distribution of the radiation for the different SRS components.

The durations of the exciting-radiation pulse and of the first SRS Stokes component were measured using
a seven-stage photomultiplier of the ÉLU-FT type, with a semitransparent oxygen—silver—cesium end-window
photocathode. The time constant of the photomultiplier was approximately 2 nsec, the integrated sensitivity of
the photocathode was 8.15 μA/lum, the spectral sensitivity region was 380-1100 nm, and the gain was $\alpha = 2.5 \cdot$
10^5. To register the signal from the ÉLU-FT we used an I2-7 nanosecond time-interval meter operating in the
single triggering regime. The oscilloscope was triggered by the investigated pulse itself. To separate the required SRS spectral component a stack of light filters was placed in front of the ÉLU-FT. The signal from the
oscilloscope screen was photographed with a "Zenit V" camera with "Jupiter-9" lens (aperture 1:2) on photographic film RF-3, on which time markers were also photographed.

CHAPTER III

WIDTH OF SPECTRAL DISTRIBUTION OF THE COMPONENTS OF SRS OF LIGHT

1. Line Contour of First SRS Stokes Component

This section presents the results of an investigation of the spectral distribution of the intensity of the
first SRS Stokes component, observed by us experimentally in media with different Kerr constants following
excitation by a laser that generates one longitudinal mode.

If the employed spectral instruments have high resolution and dispersion, the spectral distribution of the
intensity of the first SRS components of all the investigated substances, regardless of the value of the Kerr
constant, reveal a fine structure consisting of individual components. By way of illustration, Fig. 3 shows an
interferogram of the spectral distribution of the intensity of the first Stokes component of SRS of liquid nitrogen, which shows three orders of interference, at an excitation energy 0.03 J and at a scattering-layer thickness 50 mm. The same figure shows a microphotogram of the contour of this distribution in first order of

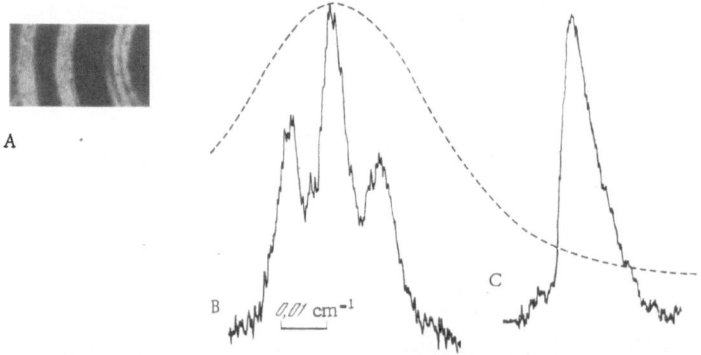

Fig. 3. Spectral distribution of the intensity of the first SRS component of liquid nitrogen. (A) Interferogram of radiation with Fabry–Perot etalon having a base 75 mm; (B) microphotogram of contour in first order of interference; (C) microphotogram of the contour of the exciting laser radiation; dashed line – contour of ordinary RS line.

interference, the line contour of the corresponding oscillation of the ordinary Raman scattering, and the line contour of the exciting radiation. It should be noted that at the given dispersion and resolution of the etalon a single line is always observed in the spectral distribution of the intensity of the exciting radiation.

Our investigations have shown that the number of components in the spectral distribution of the intensity of the first SRS Stokes component depends on the excitation conditions, and in particular on the intensity of the exciting radiation. It is known that the experimental plot of the intensity of the first SRS Stokes component against the intensity of the exciting radiation $I_{SRS}(I_0)$ has a complicated form. By way of example Fig. 4 shows the dependence of the intensity of the first SRS Stokes component of liquid nitrogen on the intensity of the exciting radiation, obtained by a number of workers [25, 60] (the so-called energy curve). At low excitation intensities this curve is nearly exponential, and when a definite value $I_0 = I_{thr}$ is reached for each substance, the SRS intensity increases in a small interval of I_0 by several orders of magnitude, followed by saturation I_{SRS}.

In the region of the curve up to the "jump" I_{SRS} we observed in the spectral distribution of the first Stokes component one or two narrow lines. With increasing excitation intensity I_0 appear additional spectral components with constant spectral width in the spectral distribution of the intensity of the first Stokes component. The spectral width of the contour of the envelope of the components of the fine structure of the first Stokes SRS component was always narrower than the width of the corresponding RS oscillation. We shall henceforth call the spectral width of the contour of the envelope with the components of the fine structure the summary spectral width of the component $\Delta\nu_{SRS}$, and the distance between the components of the structure will be designated $\delta\nu$.

2. Summary Width of Spectral Distribution of the Intensity
of the SRS Stokes Components

Since the form of the spectral distribution of the intensity of the first SRS Stokes component depends on the intensity of the exciting radiation, we measured the summary width of the first SRS Stokes component at

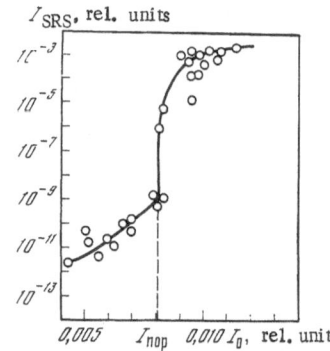

Fig. 4. Plot of the intensity of the first SRS Stokes component of liquid nitrogen against the intensity of the exciting radiation [25, 60].

TABLE 2

Substance	ν_R cm⁻¹	$\dfrac{d\sigma_\parallel}{d\Omega}\cdot 10^{+30}$, cm² · sr⁻¹	$N\cdot 10^{-22}$, cm⁻³	$g\cdot 10^3$, cm/MW theory	experiment	I_{thr}, J	l, cm
1	2	3	4	5	6	7	8
Liquid nitrogen	2330	0.185±0.055 [25]	1.698	1.80	1.4±0.6 [64]	0.012±0.001	5
Calcite	1086						
300°K		8 [61]	1.622	2.28		0.03±0.003	1.2
77°K	1086	8 [61]	1.622	5.05	4±1 [65]	0.02±0.003	1.2
Carbon disulfide	656	9.1±2.7 [25] 10.2 [62]	0.83	2.4	2.0±0.5 [66]	0.008±0.0005	5
Nitrobenzene	1345	22.3±0.4 [62] 24 [63]	0.587	3.32		0.02±0.003	20

excitation intensities I_0 up to the "jump" of I_{SRS} and passed the "jump" under conditions close to the saturation of I_{SRS}. The results were compared with the values calculated from formulas (2) and (3) (Chapter I). The data needed for the theoretical estimate of the line width of the first SRS Stokes component are given in Table 2.

The first column of the table indicates the investigated substance. In the second column of Table 2 is given the frequency of the molecular vibration, and in the third the differential RS cross sections calculated from the published data for $\lambda = 6943$ Å. The fourth column lists the number of molecules in 1 cm³, and the fifth and sixth columns give the theoretical and experimental values of the gain at the wavelength of the first SRS Stokes component. The theoretical gains were calculated from formula (1) (Chapter I). The seventh column of the table gives the threshold values of the energy of the exciting radiation for the first SRS Stokes component, which were measured by us in experiment.* The last column of the table gives the thicknesses of the scattering layers.

Table 3 gives the results of a comparison of the width of the summary distribution of the first SRS Stokes component of a number of substances with formulas (2) and (3). In the first column of the table is indicated the substance, in the second the frequency of the molecular vibration, and in the third the width of the RS line. In the last four columns of the table are given the SRS line widths in the region of saturation of the intensity of I_{SRS} to the line and in the region of the exponential growth of I_{SRS} near the sensitivity threshold, measured in experiment, as well as the values calculated theoretically.

The width of the SRS line near the threshold, given in the table, corresponds to the width of one fine-structure component (this value does not depend on the experimental conditions). In the region of saturation of the intensity of I_{SRS}, the value cited is the summary width of the first Stokes component.

*The values of I_{thr} given in the table correspond to the values of I_0 ahead of the "jump" of I_{SRS} and determine the experimental threshold of the sensitivity of installation.

TABLE 3

Substance	$\nu_R\cdot$ cm⁻¹	γ_{RS}, cm⁻¹	$\Delta\nu_{SRS}$, cm⁻¹ near the threshold experiment	theory	in the saturation region experiment	theory
Liquid nitrogen	2330	0.067 [67]	0.006 *±0.0005	0,06	0.054±0.005	0.05
Calcite						
77°K	1086	0.5 [68]	0.008† ±0.0005	0,11	0.067±0.003	0.09
300°K	1086	1.1 [13, 68]	0.02† ±0.002	0.25	0.18±0.01	0.21
300°K	2172				0.26±0.01	0.30
300°K	3257				0.28±0.01	0.36
Carbon disulfide	656	0.5 [67]	0.015† ±0.001	0.15	0.096±0.005	0.12
Nitrobenzene	1345	8 [69] 6.7 [70]	0.25† ±0.02	3,1	2.91±0.05	2.60

Note. The maximum errors in the calculation of theoretical values of $\Delta\nu_{SRS}$ (on the order of 20%) is given by the value of g.
*The width of the instrumental function is 0.002 cm⁻¹, and the line width of the exciting laser radiation is 0.015 cm⁻¹.
†Corresponds to the width of the instrumental function of the apparatus.

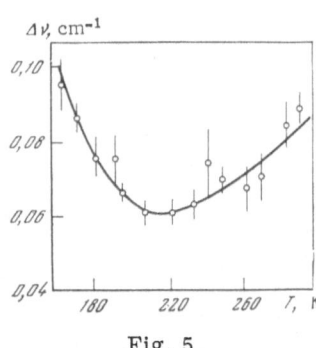
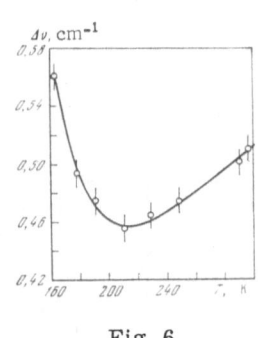
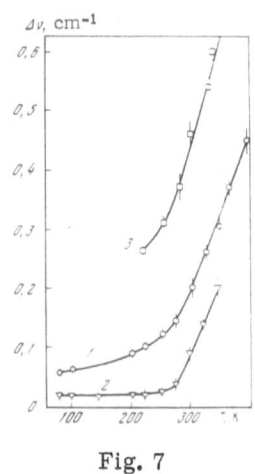

Fig. 5   Fig. 6   Fig. 7

Fig. 5. Temperature dependence of the line width of the first SRS Stokes component of carbon disulfide.

Fig. 6. Temperature dependence of the line width of the 656 cm^{-1} molecular oscillation of ordinary RS of carbon disulfide [77].

Fig. 7. Dependence of the summary width of the SRS components of calcite on the temperature. (1) First forward SRS Stokes component; (2) first backward SRS Stokes component; (3) second forward SRS Stokes component.

As seen from Table 3, at low excitation intensities the experimentally measured line width of the first SRS component is smaller by one order of magnitude than the width calculated theoretically, and is smaller by two orders of magnitude than the RS line width. When the intensity saturates, the width of the summary distribution is on the order of the calculated value. Summarizing the results, we can state that the theoretical expressions (2) and (3) do not describe the spectral widths of the SRS components. The increase of the summary width of the first SRS component with increasing excitation intensity and the very narrow width of the individual component near the threshold do not agree with the theoretical premises under which expressions (2) and (3) were obtained. It appears that the agreement between the theoretical and experimental values of $\Delta\nu_{SRS}$ in the region of saturation of I_{SRS} is for the most part accidental.

3. Investigation of the Width of the First SRS Stokes

Component of Carbon Disulfide at Various Temperatures

When the temperature of the medium changes, a substantial change takes place in the width of the RS lines with nonzero degree of depolarization. In the case of ordinary RS, as shown theoretically [71] and experimentally [72-76], this is due to the change of the average time of reorientation of the molecules with changing temperature. It was of interest to investigate the influence of the temperature on the spectral distribution of the intensity of the first Stokes component following excitation of SRS by a laser that generates one longitudinal mode. Under these conditions we investigated the width of the first SRS Stokes component of liquid carbon disulfide in the temperature interval from 165 to 300°K (the freezing point is 161.4°K).

Our investigations have shown that the width of the fine-structure component observed in the structural distribution of the first Stokes component remains constant in the entire temperature interval. A substantial change with temperature takes place in the width $\Delta\nu_{SRS}$ of the summary spectral distribution of the SRS components. Investigations of the temperature dependence of $\Delta\nu_{SRS}$ were carried out at excitation intensities corresponding to the region beyond the "jump" of I_{SRS} (see the "energy" curve, Fig. 4). The width of the summary spectral distribution in this region does not change significantly when I_0 is varied. We present below the measured widths of the first SRS Stokes component of the 656 cm^{-1} oscillation of carbon disulfide at various temperatures of the medium:

T, °K	$\Delta\nu_{SRS}$, cm^{-1}	T, °K	$\Delta\nu_{SRS}$, cm^{-1}
298	0.096 ± 0.005	223	0.058 ± 0.002
273	0.074 ± 0.005	213	0.056 ± 0.002

T, °K	$\Delta\nu_{SRS}$, cm^{-1}	T, °K	$\Delta\nu_{SRS}$, cm^{-1}
268	0.072 ± 0.004	193	0.068 ± 0.004
253	0.065 ± 0.004	173	0.087 ± 0.004
233	0.060 ± 0.003	165	0.105 ± 0.005

A plot of $\Delta\nu_{SRS}$(T) is shown in Fig. 5. It was of interest to compare the curve shown in Fig. 5 with the analogous dependence for the fully symmetrical 656 cm^{-1} oscillation of ordinary RS of carbon disulfide. These data were kindly supplied to us by Stoicheff. The temperature dependence of the line width of the 656 cm^{-1} oscillation of carbon disulfide was measured, with excitation by an argon laser, by Stoicheff and co-workers [77] in the temperature interval from 163 to 300°K. The dispersion system was a Fabry–Perot etalon with a 3 mm base. The experimental plot of γ_{RS}(T) obtained in that reference is shown in Fig. 6. As seen from Figs. 5 and 6, the width $\Delta\nu_{SRS}$ is approximately one-fifth the width of the γ_{RS} line. The waveform of the $\Delta\nu_{SRS}$ curve duplicates the waveform of the γ_{RS}(T) curve. Both curves have a minimum at the same temperature near 210°K.

4. Investigation of the Width of the SRS Components of Light at Various Temperatures of the Medium in the Forward and Backward Directions

The temperature dependence of the summary width of the SRS components was investigated also for single-crystal calcite. The width $\Delta\nu_{SRS}$ of the first Stokes component was measured simultaneously both in the direction of propagation of the exciting radiation ("forward") and in the opposite direction ("backward"). We measured also the spectral width of the second Stokes component forward. Table 4 gives the results of these measurements.

The first column of the table indicates the temperature, the second and third indicate $\Delta\nu_{SRS}$ of the first Stokes component forward and backward, and the fourth gives $\Delta\nu_{SRS}$ of the second Stokes component forward. Plots of $\Delta\nu_{SRS}$(T) corresponding to the tabulated data are shown in Fig. 7. From Table 4 and Fig. 7 it is seen that $\Delta\nu_{SRS}$ varies significantly in the temperature range from 77 to 400°K. The width $\Delta\nu_{SRS}$ of the first Stokes component, both forward and backward, increases by one order of magnitude, while that of the second Stokes component increases by approximately four times. The shape of the $\Delta\nu_{SRS}$(T) curves is close to exponential.

The line width of the fully symmetrical oscillation of ordinary RS when excited by an argon laser was measured by Park [68] at 77 and 300°K. These data are given in the last column of Table 4. When the calcite temperature is changed from 77 to 300°K the RS line narrows down by a factor 2.2, and the line width of the first SRS Stokes component decreases by a factor 2.5.

To establish the extent to which the difference between the widths $\Delta\nu_{SRS}$ in the forward and backward directions is typical, we measured the width of the spectral distribution of the first Stokes component of a number of other substances in these two directions. The SRS spectra forward and backward were registered simultaneously. These experimental results are given in Table 5.

TABLE 4

T, °K	$\Delta\nu_{SRS}$, cm^{-1}		second Stokes component "forward"	γ_{RS}, cm^{-1} [13, 68]
	first Stokes component			
	forward	backward		
77	0.067 ± 0.003	0.020 ± 0.002		0.5
213	0.098 ± 0.003	0.022 ± 0.002		
220	0.110 ± 0.005	0.022 ± 0.002	0.27 ± 0.02	
233	0.12 ± 0.01	0.022 ± 0.002		
243	0.125 ± 0.005	0.022 ± 0.002	0.30 ± 0.02	
253	0.14 ± 0.01	0.030 ± 0.002		
273		0.035 ± 0.003	0.37 ± 0.03	
283	0.16 ± 0.01	0.040 ± 0.004		
300	0.18 ± 0.01	0.060 ± 0.004	0.48 ± 0.03	1.1
323	0.24 ± 0.01	0.095 ± 0.005	0.55 ± 0.05	
338	0.29 ± 0.01	0.20 ± 0.02	0.67 ± 0.05	
343	0.32 ± 0.02		0.73 ± 0.05	
363	0.37 ± 0.02			
383	0.45 ± 0.03			

TABLE 5

Substance	$\Delta\nu_{SRS}$, cm⁻¹	
	forward	backward
Liquid nitrogen	0.054±0.005	0.043±0.004
Nitrobenzene	2.92±0.005	1.81±0.005
Carbon disulfide	0.096±0.005	0.032±0.003

It is seen from Table 5 that the width of the summary spectral distribution of the intensity of the first Stokes component of the SRS component in the backward direction is smaller for all the investigated substances than in the forward direction.

The experimental results lead to the following conclusions:

a. SRS of light in all the investigated substances has a complicated spectral distribution of the intensity, consisting of individual components. It is shown that neither the width of the individual component of the structure nor the spectral width of the contour of the envelope of the components can be satisfactorily described by a theoretical relation that takes into account the narrowing of the spectral width of the RS line when radiation passes through a nonlinear medium.

b. The width of the summary spectral distribution of the intensity of the SRS components depends on the temperature of the medium. With calcite and carbon disulfide as examples, it was shown that the width of the first SRS Stokes component changes with changing temperature in accordance with the change of the width of the RS line.

c. The summary width of the spectral distribution of the intensity of the SRS components in the direction of propagation of the exciting radiation is always larger than the width in the opposite direction.

CHAPTER IV

FINE STRUCTURE IN THE SPECTRAL DISTRIBUTION OF THE INTENSITY
OF THE COMPONENTS OF SRS OF LIGHT

1. Fine Structure of SRS Components of Liquid Nitrogen,

Calcite, Carbon Disulfide, and Nitrobenzene

The spectral distribution of the intensity of the SRS components in liquid nitrogen, calcite, carbon disulfide, and nitrobenzene, as shown in Chapter III, consists of a number of components which can be observed if the resolution of the spectral instrument is sufficiently high. By way of illustration, Fig. 8 shows a spectral distribution of the intensity of the exciting radiation and of the SRS components of single-crystal calcite, obtained with the aid of a Fabry–Perot etalon with an 8-mm base. A fine structure was observed in all the Stokes components. The spectral distribution of the intensity of the exciting laser radiation passing through the medium remains smooth at a given dispersion and resolution of the etalon. Figure 9 shows microphotograms of the contour of the first and second SRS Stokes components of calcite and the line contour of the exciting radiation. The investigations show that a fine structure is observed also in the SRS anti-Stokes components.

Figure 10 shows a microphotogram of the contour of the first SRS Stokes component of carbon disulfide, obtained in the same optical system with an etalon having a base of 14 mm and at an excitation energy close to I_{thr}. The spectral distribution of the intensity shows a fine structure whose distances between the components, just as for calcite and for liquid nitrogen, do not correspond to the distance between the SMBS components and the modes of the laser. The summary width of the spectral distribution of the intensity is less in this case than the width of the line of the ordinary RS. The spectrum of the first Stokes component shows two fine-structure components, and when the excitation intensity is increased the number of the fine-structure components and the summary width of the spectral distribution increase up to an excitation power 4.5 MW. At a power close to 4.5 MW, large spectral broadenings of the first SRS components (on the order of hundreds of reciprocal centimeters) appear, and the contour assumes a more complicated form.

The spectral width of an individual component of the fine structure is smaller by almost two orders of magnitude than the RS line width. In the case of calcite, and also carbon disulfide, it amounts to 0.008-0.02 cm⁻¹, i.e., it is on the order of the width of the instrumental function of the etalon; for liquid nitrogen it amounts

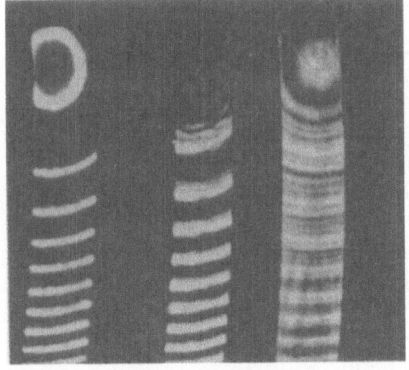 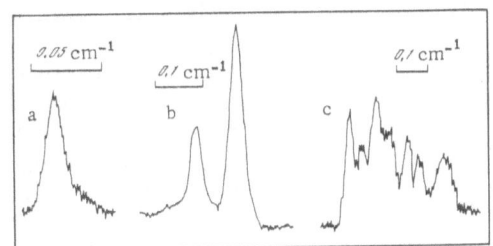

Fig. 8   Fig. 9

Fig. 8. Spectral distribution of the intensity of the exciting radiation passing through the medium and of the SRS components of single-crystal calcite.

Fig. 9. Microphotogram of the distribution of the intensity at 300°K: (a) of the exciting radiation passing through calcite; (b) of the first and (c) of the second SRS Stokes components; the RS line has a dispersion contour of width 1.1 cm^{-1}.

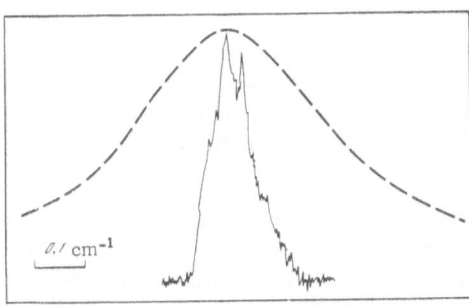 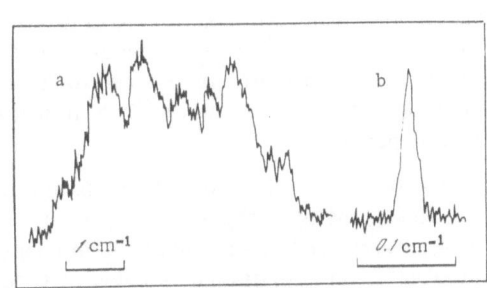

Fig. 10   Fig. 11

Fig. 10. Microphotogram of the contour of the first SRS Stokes component of carbon disulfide at an excitation intensity near I_{thr}. Temperature 300°K, dashed curve — contour of RS line. Fabry—Perot etalon with 14 mm base.

Fig. 11. Microphotogram of the contour of the line of the first SRS component of nitrobenzene (a) and of the exciting radiation passing through the medium (b). Fabry—Perot etalon with 1.3 mm base.

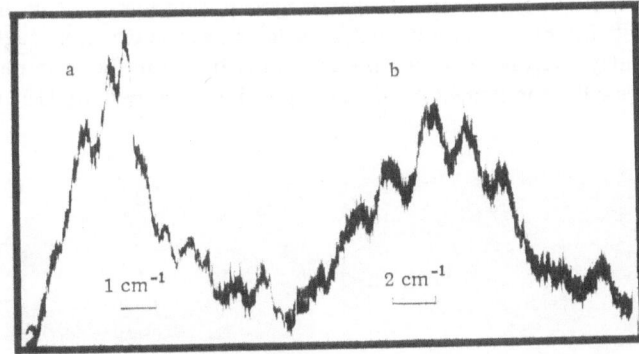

Fig. 12. Microphotogram of the intensity distribution of the first Stokes (a) and of the second Stokes (b) components of SRS of nitrobenzene. Fabry—Perot etalon with 0.6 mm base.

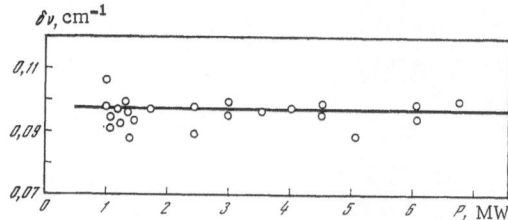

Fig. 13. Dependence of the distances between the components of the fine structure of the first SRS Stokes component in liquid nitrogen on the power of the exciting radiation (in the forward direction).

to 0.006 cm^{-1} at an instrumental-function width 0.002 cm^{-1}. The width of the line of the exciting laser radiation is 0.015 cm^{-1}.

A fine structure is observed also in the spectral distribution of the intensity of the first SRS Stokes component of nitrobenzene, although owing to the unfavorable ratio of the large summary line width and the small distance between the components of the structure, its registration is a rather complicated matter. Microphotograms of the contour of the first SRS Stokes component of nitrobenzene and of the exciting radiation passing through the medium are shown in Fig. 11. A fine structure is observed also in the spectral distribution of the intensity of the first SRS Stokes component, and the intensity distribution turns out to be more complicated than in the case of the first Stokes component (see Fig. 12).

A characteristic feature of the fine structure observed by us in the spectral distribution of the intensity of the SRS components is a certain randomness and the appearance of the number and sequence of the intensity of the components of the structure from flash to flash. If we consider the cross section of the beam of an individual SRS component, then we can observe in different parts of this beam somewhat differing numbers of fine-structure components.

The singularities of the fine structure observed by us were investigated in greatest detail in the spectral distributions of the intensity of the SRS components in liquid nitrogen and calcite. Table 6 lists the summary widths of two SRS Stokes and one anti-Stokes components of calcite in the forward direction, the number of the observed components, and the distances between them at a temperature 300°K.

Compared with the first Stokes component, the number of the fine-structure components is larger, and the distances between them are smaller for the first anti-Stokes and for the second Stokes components.

We have also investigated the dependence of the width of the spectral distribution and the numbers of the fine-structure components of the first Stokes component of SRS of liquid nitrogen on the intensity of the exciting radiation. The results of these measurements are given in Table 7.

With increasing intensity of the exciting radiation, the number of the fine-structure components and the summary width of the spectral distribution of the first SRS Stokes component of nitrogen increase. The width of an individual component of the structure remains unchanged in this case. At an intensity 65 MW/cm^2, saturation of the intensity of the first SRS Stokes component sets in. In the saturation region, the spectral distribution does not change significantly. These results agree with the data for the summary spectral distribution of the intensity of the first Stokes SRS component of stilbene powder obtained in [18] (where the fine structure was not observed).

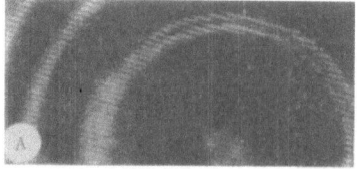 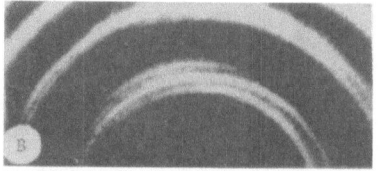

Fig. 14. Spectral distribution of the intensity of the first SRS component of liquid nitrogen in the backward (A) and forward (B) directions. The photographs were obtained simultaneously; Fabry‡Perot etalon with 75 mm base.

SRS component	Summary width of spectral distribution, cm^{-1}	Number of component	Splitting, cm^{-1}
1St	0.18	3—4	0.05
2St	0.26	8—9	0.03
1aSt	0.18	6—7	0.03—0.02

TABLE 6

I_{exc}, MW/cm^2	Number of fine-structure components	Summary width of spectral distribution, cm^{-1}
10—13	1—3	0.006—0.010
13—38	3—4	0.016—0.019
55	5	0.022
65	5—9	0.054

TABLE 7

Investigations have shown that the distances between the fine-structure components ($\delta\nu$) do not depend on the excitation energy. By way of example, Fig. 13 shows the dependence of the distance between the fine-structure components of the first SRS Stokes component of liquid nitrogen in the forward direction on the exciting-radiation power. It is seen from the figure that the distance between the components remains the same.

Simultaneous observation of SRS in the forward and backward directions has shown that the number of components in the spectral distribution of the first Stokes component is different in the two directions. One of the typical interferograms is shown in Fig. 14. The number of components and the magnitude of the splitting varied somewhat from experiment to experiment, but the asymmetry in these two directions was as a rule preserved.

2. Quasiperiodicity of Spectral Distribution of the Intensity of the First SRS Stokes Component

In the study of the spectral distribution of the intensity of the SRS components, attention is attracted by the almost equidistant arrangement of the components of the fine structure. To estimate how regular the observed structure is, we have measured the distances between the components of the fine structure on 30 interferograms for liquid nitrogen and on more than 80 for calcite.

For each interferogram we determined the average distance between the fine-structure component

$$\delta^{(m)}\nu = \frac{1}{n}\sum_{i=1}^{n}\delta_i\nu,$$

where n is the number of the fine-structure intervals; $\delta_i\nu$ is the i-th fine-structure interval.

It turned out that at any temperature the substance has a distinctive characteristic average distance between the fine-structure components $\delta\nu$. We have also determined the mean-squared deviation from the average distance between components:

$$\delta(\delta^{(m)}\nu) = \frac{1}{m}\sqrt{\sum_{i=1}^{m}(\delta\nu - \delta^{(i)}\nu)^2},$$

$$\delta\nu = \frac{1}{m}\sum_{i=1}^{m}\delta^{(i)}\nu,$$

where m is the number of photographs.

The mean-squared deviation is less than 10%. This means that within the limits of the measurement errors (about 10%) the average distance between the fine-structure component is preserved from flash to flash (for a given temperature of the substance).

To estimate the regularity of the fine structure, we determined the fraction of photographs on which the observed structure was regular within the limits of the measurement errors (10%). To this end we calculated for each photograph the variance

$$\sigma_m = \frac{1}{n}\sqrt{\sum_{i=1}^{n}|(\delta^{(m)}\nu)^2 - (\delta_i\nu)^2|}.$$

The regularity of the arrangement of the components of the fine structure is characterized by the ratio of the variance to the mean distance between the components. The fraction of photographs for which this ratio was less than or equal to the measurement error turned out to be 70% in our case.

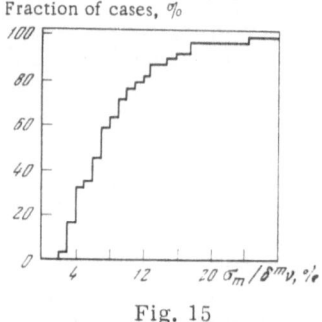

Fraction of cases, %

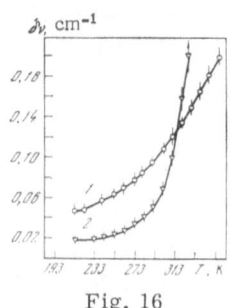

$\delta\nu$, cm^{-1}

Fig. 15   Fig. 16

Fig. 15. Dependence of the fraction of cases on the ratio of the variance to the mean distances between the fine-structure components.

Fig. 16. Dependence of the distance between the fine-structure components of the first SRS Stokes component of a calcite crystal on the temperature: (1) forward; (2) backward.

By way of example, Fig. 15 shows the dependence of the fraction of cases on the ratio of the variance to the mean distance between the fine-structure components $(\sigma_m/\delta^{(m)}\nu)$ for the first Stokes component of single-crystal calcite. Four fine-structure components were observed on the processed photographs.

Thus, for a given temperature of the scattering medium, in 70% of the cases the fine structure observed in the field of the first Stokes component is regular within the limits of the measurement errors, and the average distance between the components remains the same from flash to flash.

3. Influence of the Temperature of the Scattering Medium on the Distance between the Fine-Structure Components of the First SRS Stokes Component of Single-Crystal Calcite

When the temperature of the calcite crystal is raised from 77 to 400°K the width of an individual fine-structure component remained practically unchanged, whereas the distance between the components increased noticeably. Table 8 gives the results of the measurements of the average distance between the components of the fine structure of the first SRS component of calcite at various crystal temperatures. The SRS spectra were registered in the forward and backward directions simultaneously.

Plots of $\delta\nu(T)$ corresponding to the tabulated data are shown in Fig. 16.

As seen from the table and from the figure, when the temperature is increased from 77 to 400°K the distances between the fine-structure components increase by approximately five times, considerably in excess of the measurement error (10%).

The average distances between the fine-structure components at equal temperatures are always different in the forward and backward directions, as are also the numbers of the fine-structure components (a similar picture was observed for liquid nitrogen (see, e.g., Fig. 14).

TABLE 8

T, K	$\delta\nu$ SRS, cm^{-1}		T, K	$\delta\nu$ SRS, cm^{-1}	
	forward	backward		forward	backward
77	0.037±0.004	0.017±0.002	300	0.12±0.01	0.050±0.005
213	0.046±0.004	0.020±0.002	323	0.13±0.03	0.100±0.007
233	0.055±0.004	0.022±0.002	338	0.14±0.01	0.20±0.02
243	0.063±0.005	0.022±0.002	343	0.15±0.01	
253	0.070±0.005	0.032±0.002	363	0.18±0.01	
263	0.080±0.006	0.032±0.002	383	0.22±0.02	
273	0.108±0.008	0.037±0.003			

Thus, in all the investigated substances, a fine structure is observed in the spectral distribution of the SRS-component intensity, and has the following features:

a. The width of an individual component of the structure is equal to the instrumental width of the spectral instrument and is smaller by two orders of magnitude than the width of the RS line. In the case of liquid nitrogen, the width of an individual structure component is three times the width of the instrumental function and one-third the width of the exciting-radiation line.

b. In 70% of the cases the fine-structure components are almost equidistant within the limits of the line width of the ordinary RS, and the distance between the components at a given temperature of the substance remains the same from flash to flash.

c. The distances between the fine-structure components do not depend on the energy of the exciting radiation.

d. When the temperature of the scattering medium increases, the distances between the components increase.

e. The number of components of the structure depend on the energy of the exciting radiation and increases with increasing energy.

f. The numbers of the spectral components are different in the forward and backward directions.

g. The distances between the fine-structure forward components are always larger than the backward ones in the temperature interval from 77 to 323°K.

CHAPTER V

SELF-FOCUSING, LARGE SPECTRAL BROADENINGS, AND

SPATIAL DISTRIBUTION OF THE INTENSITY OF THE

COMPONENTS OF SRS OF LIGHT

1. Self-Focusing of Light and Spectral Distribution of the

Intensity of the SRS Components in Carbon Disulfide at

Various Excitation Energies

Investigations of the evolution of self-focusing of SRS of light and of the spectral distribution of the intensity in the self-focusing region as functions of the power of the exciting radiation were carried out in liquid nitrogen, calcite, nitrobenzene, and carbon disulfide. The picture of the spectral distribution of the intensity in the cross sections of the beams of the exciting radiation and of the SRS components was observed in the interior of the medium in planes located at various distances from the entrance or exit face of the sample.

The distribution of the SRS component intensity of carbon disulfide in a plane near the exit face of the cell, at various excitation powers, is shown in Fig. 17. The figure shows the cross sections of the beams of the exciting radiation passing through the medium, and of four Stokes and three anti-Stokes SRS components. No self-focusing was observed in the section of the exciting beam at the exit face of the cell. This is due, as shown in [12], to the large coefficient of conversion of the exciting radiation into SRS of light.

Near the SRS threshold (power 0.45 MW), one self-focusing dot is observed in the section of the beam of the first Stokes component; this dot corresponds to a ring in the field of the second Stokes component. With increasing excitation power many rings of various diameters appear in the sections of the beams of the Stokes and anti-Stokes components, including systems of concentric rings. The number of self-focusing dots in the field of the first Stokes component, as well as the number and diameter of the rings of the components of higher order in the case of observation in the forward direction, depended on the power of the exciting radiation. With increasing excitation power, the number of self-focusing dots in the section of the beam of the first Stokes component increases, and their dimensions decrease. Accordingly, the number of the rings of higher Stokes and anti-Stokes component increases and their diameter decreases. Figures 18 and 19 show plots of the diameters of the dots in the field of the first SRS Stokes component and of the diameters of the rings in the field of the second SRS Stokes component against the power of the exciting radiation for carbon disulfide 20 and 50 mm thick. As seen from the figures, at a definite excitation power and a given scattering layer, the rings in the sections of the beams of the higher-order components give way to dots. With further increase of the excitation

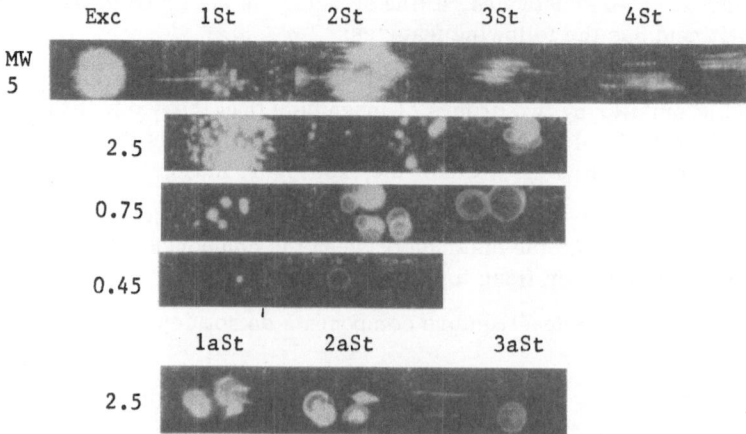

Fig. 17. Distribution of the intensity in the sections of the beams of the exciting radiation passing through the medium, and of the four Stokes and three anti-Stokes SRS components of carbon disulfide, at various values of the exciting-radiation power (in the forward direction).

power each self-focusing dot in the SRS-component beam section is spread tens and hundreds of reciprocal centimeters over the spectrum.

Figure 20 shows a microphotogram of the spectral distribution of the intensity of the first SRS Stokes component in the saturation region I_{SRS} (see Fig. 17). A microphotogram of the spectral distribution of the intensity of the central and most intense part of this broadening is shown under the same conditions in Fig. 20b. The intensive part of the contour has a complicated structure consisting of individual components with spectral widths approximately 0.1 cm^{-1} (the width of the instrumental function). The distance between the components is 0.21 cm^{-1}, which corresponds to the distance between the SMBS components in liquid carbon disulfide. The total width of the intensive part of the contour amounts to 1.6 cm^{-1}, i.e., it exceeds by three times the width of the line of the fully symmetrical 656 cm^{-1} vibration of carbon disulfide in ordinary RS. Less intensive wings produce a continuous background on the interferogram.

Figure 21 shows graphically the average value of the width of the spectral distribution of the self-focusing points of the SRS components at different powers of the exciting radiation. The numerical values of the width of the spectral distribution for the Stokes components are given in Table 9.

The spectral broadening is produced at a certain threshold value of the power of the exciting radiation (4-4.5 MW at a scattering-layer thickness 50 mm) practically simultaneously on all the observed components. For the first Stokes component the broadening extends predominantly into the anti-Stokes region, while for the higher-order components it extends into the Stokes region.

Distinct self-focusing dots are observed in the entire interval of exciting-radiation power in the backward direction in the sections of the SMBS beams of the exciting radiation and of the SRS components near the entrance face of the cell. No rings were observed in the fields of the second and third Stokes components. Figure 22 shows the distribution of the intensity in the sections of the SMBS beams and of the first and second Stokes components of SRS at various exciting-radiation powers.

TABLE 9

P, MW	Width of spectral distribution, Å				P, MW	Width of spectral distribution, Å			
	1St	2St	3St	4St		1St	2St	3St	4St
4.3	19.2	12.8	38.4	12.8	5.65	—	38.4	—	—
4.4	—	—	57.6	102.4	5.9	19.2	51.2	19.2	—
4.5	25.6	38.4	51.2	—	6.1	12.8	25.6	38.4	25.6
4.55	38.4	63.8	32	—	6.6	19.2	32	51.2	44.8
4.6	95.8	32	—	—	6.75	—	38.4	64.8	—

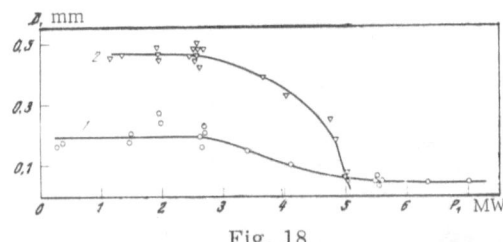
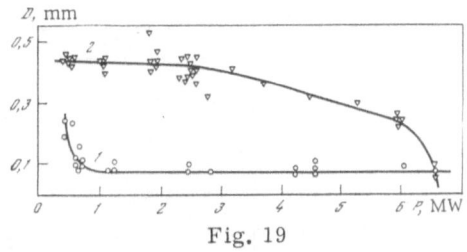

Fig. 18   Fig. 19

Fig. 18. Diameter of the self-focusing dots in the field of the first SRS component (1) and diameters of the rings in the field of the second Stokes component (2) of carbon disulfide in the forward direction as functions of the power of the exciting radiation at a layer thickness 20 mm.

Fig. 19. Diameters of self-focusing dots in the field of the SRS first Stokes component (1) and diameters of the rings in the field of the second Stokes component (2) of carbon disulfide in the forward direction as functions of the power of the exciting radiation at a layer thickness 50 mm.

At a definite excitation power, large spectral broadenings appear at the self-focusing dots of the second SRS Stokes component in the backward direction, just as in the forward direction, but the thresholds of the appearance of the broadenings in these directions are different. The threshold of the appearance of the backward broadening is lower than that of the forward broadening: at a layer thickness 50 mm, the threshold for the appearance of broadenings in the forward direction is 4-4.5 MW, as against 2-2.5 MW in the backward direction. It should also be noted that in the backward direction interference rings are observed around the self-focusing dots of the first Stokes components.

2. Self-Focusing and Spectral Distribution of the Intensity of the SRS Components in Nitrobenzene, Calcite, and Liquid Nitrogen

We investigated the distribution of the intensity in the sections of the beams of the SRS components of nitrobenzene, calcite, and liquid nitrogen. Self-focusing dots of the first Stokes component and rings of higher-order components were observed for all the indicated substances in the cross sections of the SRS components in planes near the exit face of the sample. By way of example, Fig. 23 shows the spatial distribution of the first and second SRS Stokes components of calcite near the exit face of the sample. With increasing excitation power, self-focusing dots appear simultaneously for all components.

In nitrobenzene at high excitation powers (approximately 11 MW) at the self-focusing dots of the second Stokes component, spectral broadenings up to 100 cm^{-1} are observed. In liquid nitrogen and calcite no spectral broadenings of the SRS components were observed in the entire interval of values of the power of the exciting radiation.

It should be noted that in nitrobenzene, calcite, and liquid nitrogen the experimentally observed pictures of the change in the spatial and spectral distribution of the intensity of the SRS components as a function

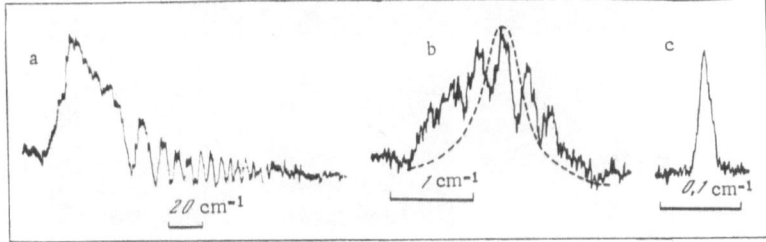

Fig. 20. Microphotogram of the contour of the first SRS Stokes component (a, b) and of the exciting radiation passing through carbon disulfide (c). (a) STÉ-1 spectrograph, dispersion 12.8 Å/mm; (b, c) Fabry—Perot etalon with base 1 mm; dashed curves — contour of RS line.

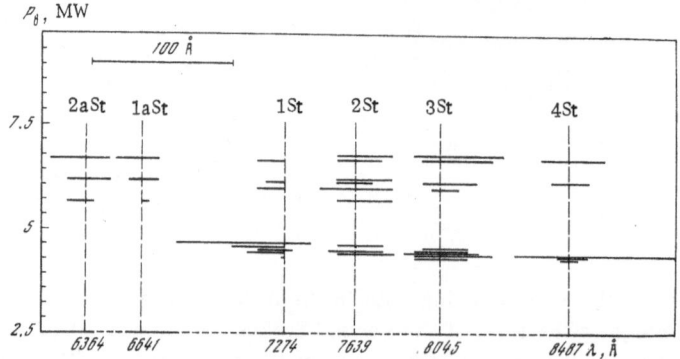

Fig. 21. Average value of the spectral width at the self-focusing points of the SRS components of carbon disulfide at various excitation powers.

of the exciting-radiation power are less distinctly pronounced than in carbon disulfide. As shown in the preceding section, the decrease of the diameters of the rings of the higher Stokes and anti-Stokes components and the appearance of spectral broadenings in carbon disulfide could be traced quite reliably in a large number of components, and these changes took place in a wide interval of excitation-power values (0.45-5 MW). This is apparently due to the relatively low value of the vibrational quantum (656 cm^{-1}), to the large dispersion, and to the conversion coefficient of the exciting radiation and of the SRS light in carbon disulfide. In the remaining substances investigated by us the rings of the higher-order components are replaced by self-focusing following a small change in the excitation power (1 MW).

3. Spatial Distribution of the Intensity of the SRS Components

In the preceding section we described the results of experiments that made it possible to observe the distribution of the intensity of the SRS components in a plane inside the medium. To estimate the aperture angles of the radiation cones, which produce rings in the beam sections inside the medium, the intensity distributions in planes at different depths inside the cell were projected in succession onto the plane of the STÉ-1 slit. The excitation energy was constant in this case and close to the threshold value. On the plane of the photographic plate, several rings were obtained for a given SRS component (from 1 to 4), with diameters that hardly differed. In the plane of the STÉ-1 slit, we projected the planes inside the cell, with distances between them 1.5-2 mm. We measured on the photographic plate the diameters of the ring and measured their dependence on the position of the observation plane.

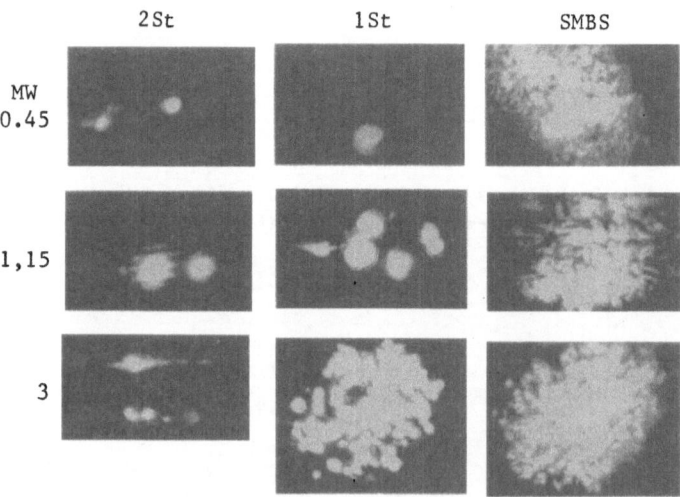

Fig. 22. Distribution of the intensity in the cross sections of the beams of SMBS and of the first and second SRS Stokes components of carbon disulfide at various values of the excitation power in the backward direction.

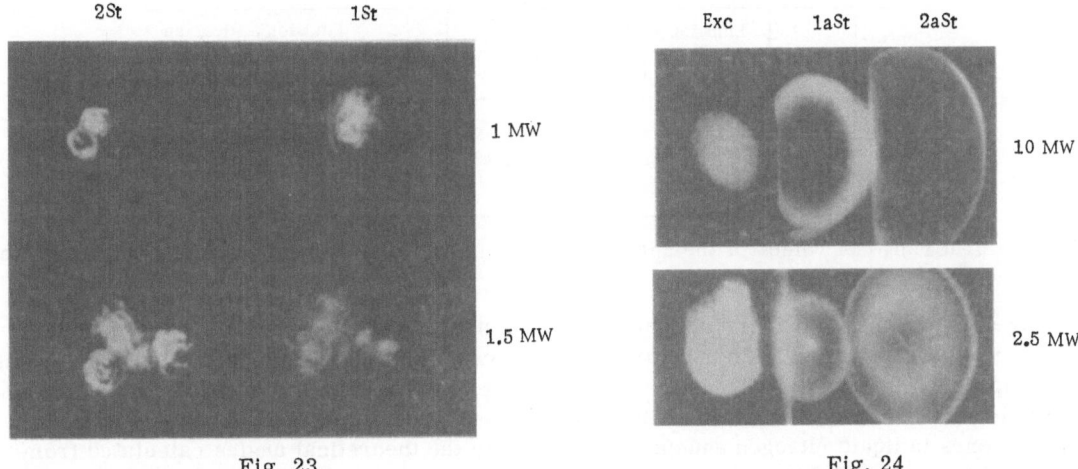

Fig. 23   Fig. 24

Fig. 23. Intensity distribution in the cross sections of the beams of the first and second SRS Stokes components of calcites at various excitation powers.

Fig. 24. Angular distribution of the intensity of the anti-Stokes SRS components of carbon disulfide at various excitation powers and at a layer thickness 50 mm.

It is known that the picture of the angular distribution of the intensity of the SRS components can be observed in the focal plane of a converging lens. We have investigated the angular distribution of the intensity of four Stokes and two anti-Stokes components of SRS of carbon disulfide. The investigations were made at different values of the excitation power. The picture of the angular distribution of the components, just as the picture of the spectral distribution of the intensity, depended on the exciting-radiation power. The distribution of the intensity of the components and the focal plane of the lens, obtained with a "Huet" three-prism spectrograph, is shown in Fig. 24. The emission angles observed in the angular distribution of the SRS of carbon disulfide are given in Table 10. All the angles were measured in air.

Excitation near the threshold revealed in the focal plane of the lens a system of low-intensity rings, with maximum intensity at the center; this system corresponds to the emission of higher Stokes and anti-Stokes SRS components. With increasing excitation power, an intense emission ring whose angle width is somewhat larger (by 2-3 times) than the angle width of the low-intensities appears in the angular distribution of the components. The power of the exciting radiation, at which intense rings appear in the emission of the high-order components, corresponds to the appearance of self-focusing points in the sections of the SRS components near the exit face of the cell and to large broadenings at the self-focusing points. With further increase of the excitation power the low-intensity thin rings drop out, as do also the maxima of the emission at the centers of the anti-Stokes components. For the Stokes component, the emission intensity on the axis increases.

The diameter of the low-intensity rings varies somewhat from flash to flash. In addition to thin sharp rings there is also observed radiation that fills the space bounded between conical surfaces. The diameter of the intense rings remains the same from flash. The angles of the observed emission cones do not depend on the excitation geometry. When the spherical lens that focuses the radiation into the medium is replaced by a cylindrical lens, the picture of the angular distribution of the components remains unchanged. The measurements performed in carbon disulfide have shown that the angle obtained in the case of measurements inside the

TABLE 10

Component	λ, Å	Emission angles, rad				Component	λ, Å	Emission angles, rad			
2aSt	6364	—	0.0423	0.061	0.0713 *	3St	8045	0.034	0.052	0.064	—
1aSt	6641	0.032	0.047 *	—	—	4St	8487	0.022	0.036	0.046	—
2St	7639	0.032	0.036 *	0.047	—						

*The values of the angles correspond to intense emission cones. The errors in the values of the angles do not exceed 10%.

TABLE 11

Component	λ, Å	Emission angles, rad		Component	λ, Å	Emission angles, rad	
		theory	experiment			theory	experiment
2 St	10264	—	0.034±0.001	3 aSt	4670	0.054	0.055±0.001
1 aSt	5976	0.018	0.0191±0.0005	4 aSt	4320	0.072	0.068±0.007
2 aSt	5246	0.039	0.0385±0.0005				

medium lies within the range of values of the angles obtained from measurements in the angular distribution outside the cell with the medium.

The study of the distribution of the intensity in the beam cross sections of the first SRS Stokes components in single-crystal calcite inside the medium, near the exit window, has shown that for higher Stokes and anti-Stokes components near the thresholds the self-focusing points in the field of the first Stokes component correspond to rings in the field of the second Stokes component. Under condition of SRS saturations, the angles of the emission cones in liquid nitrogen and calcite agree with the theoretical angles calculated from the phase-synchronism conditions [19]. Table 11 gives the theoretical and experimental emission angles obtained by us earlier [78] for the SRS components of liquid nitrogen. The angles were measured in air. The theoretical values of the angles were calculated from formulas given in the monograph of Lugovoi [19] for class-I SRS.

It is seen from the table that the measured values of the angles (θ_m) are connected by the simple relation [19]

$$\theta_m \approx m\theta_1.$$

Similar measurements made by us for calcite single crystal in the SRS intensity saturation region agree with the values obtained by others. It is easy to note that at small powers of the exciting radiation in liquid nitrogen and calcite, just as in carbon disulfide, emission of anti-Stokes components of SRS, directed along the axis of the exciting radiation, is observed [48, 78].

Thus, the investigations reported demonstrate the following:

a. When SRS is observed in planes inside the medium in the direction of propagation of the exciting radiation (forward), agreement is observed, for all the investigated media, between the self-focusing points of the first Stokes SRS component and the centers of the rings of the higher and anti-Stokes components.

b. In the backward direction, no rings are observed in the cross sections of the higher Stokes and anti-Stokes SRS component rings.

c. In substances with large Kerr constants, at a certain definite value of the exciting-radiation power, which is a characteristic of the given substance and of the given thickness of the scattering layer, self-focusing dots with large spectral broadenings are observed in the beam cross sections of all the SRS components.

d. The threshold of the onset of spectral broadenings in the forward direction is lower than that in the backward direction.

e. No explicit dependence of the broadening on the exciting-radiation power is observed.

f. In the angular distribution of the radiation outside the medium, a complicated picture is observed for carbon disulfide: near the threshold, a system of low-intensity rings is observed and varies slightly from flash to flash; in the region of powers corresponding to the appearance of large broadenings in the spectrum, intense rings appear in the angular distribution and have a width two or three times larger than the angle width of the low-intensity rings; the intense rings are well reproducible from flash to flash.

g. When the SRS intensity saturates, the angular distribution of each SRS reveals only one intense radiation cone in all the investigated substances.

CHAPTER VI

DISCUSSION OF RESULTS

Our experimental investigations have shown that for all substances the observed contour of the SRS components represents a complicated spectral distribution consisting of individual components. The spectral width of an individual component of the structure, as well as the spectral width of the contour of the envelope of the components as shown in Chapter III, cannot be described by the relations deduced from the notion of narrowing of the line γ_{RS} when the scattered light passes through a nonlinear medium. It follows from the theoretical premises that the SRS line narrows down from the threshold up to the region of saturation of the SRS intensity [see Eq. (2) of Chapter I]. At saturation, the line width remains constant [see Eq. (3) of Chapter I] and does not depend on the excitation energy or on the thickness of the scattering layer. The experimental results obtained in our earlier study [79] and by others [18, 54] contradict these concepts. The SRS lines have near the threshold a width smaller by one order of magnitude than expected from the theory (see Table 3, Chapter III, Section 2). With increasing intensity of the exciting radiation, the SRS line width does not decrease but increases. This picture is observed in liquids (liquid nitrogen [79]), single crystals (calcite [54, 79]), and powders (stilbene [18]). The increase of the width of the summary contour is due to the appearance of additional fine-structure components with constant spectral width.

The spectral width of an individual fine-structure component does not depend on the conditions of the SRS excitation, other conditions being constant, while the width of the summary contour depends on the temperature of the scattering medium.

It is known that the line width of ordinary RS with nonzero degree of polarization changes significantly when the temperature of the scattering medium changes [72-76]. The temperature dependence of the broadening of the depolarized RS lines was explained theoretically by Sobel'man [71]. This theory received an experimental confirmation in the papers of Rakov [74], Sokolovskaya [73], Rakov [75], and others. The broadening of the RS line with nonzero degree of polarization ρ is due to modulation of the scattered light when the anisotropic molecule becomes reoriented in the liquid or in the unit cell of the crystal because of the Brownian rotational motion.

If the degree of polarization is $\rho = 6/7$, the entire broadening is determined completely by $1/\tau_r$, where

$$\tau_r = {}^4\!/_3\, \pi \frac{a^3}{kT}\, \eta \exp\!\left(\frac{U_{\text{act}}}{kT}\right),$$

η is the macroscopic viscosity coefficient, a is the radius of the molecule, U_{act} is the activation energy of the molecule, and k is the Boltzmann constant.

τ_r depends on the activation energy, on the viscosity of the liquid, and on the temperature of the medium. In this connection, with decrease in temperature the line width of the ordinary RS decreases. A comparison of the results of the experiments on the measurement of the temperature dependence of the SRS line width of calcite with the temperature dependence of the width of the ordinary SR has shown that the experimental curves of these dependences are similar in form.

In carbon disulfide there is a minimum on the plot of the temperature dependence of the line width, and its position for RS and SRS is the same (at a temperature close to 210°K). It should be noted that in the investigation of the temperature dependence of the RS line width of carbon disulfide, using a mercury lamp as the excitation source, no minimum was observed in the temperature dependence. The reason is the large width of the exciting radiation and the narrow RS line. Recently, a peculiar course of the temperature dependence of the line width of the 656 cm^{-1} oscillation was observed by Stoicheff [77] in excitation of RS by an argon laser.

The appearance of a minimum on the temperature dependence of the SRS and RS line widths is quite unexpected. An explanation of the unusual temperature dependence of the line width in carbon disulfide was proposed by Alekseev and Sobel'man [80].

Two processes contribute to the temperature dependence of the broadening of polarized RS lines:

1. The already considered reorientational motion of the molecules, which increases the line width with increasing temperature (γ_{RS}).

2. The broadening due to the phase randomization of the molecular vibration ($\tilde{\gamma}_{RS}$).

Let us examine in greater detail the second mechanism. In liquids the frequency of the molecular vibration can be written in the form $\omega = \omega_0 + \Delta\omega(t)$, where ω_0 is the frequency of the molecular vibration and $\Delta\omega(t)$ is the small increment to the frequency and varies randomly with time.

In the general case, using the premises of Fourier analysis, the RS line intensity $I(\omega)$ can be written in the form

$$I(\omega) = \frac{1}{\pi} \operatorname{Re} \int_0^\infty \Phi(\tau)\, e^{-i\omega\tau}\, d\tau, \qquad (11)$$

$$\Phi(\tau) = \langle e^{i\psi(\tau)} \rangle, \tag{12}$$

$$\psi(\tau) = \int_\tau \Delta\omega(t)\, dt = \langle \Delta\omega \rangle \tau - \beta(\tau), \tag{13}$$

$$\beta(\tau) = \int_\tau [\Delta\omega(t) - \langle \Delta\omega \rangle]\, dt, \tag{14}$$

where $\langle \Delta\omega \rangle$ is the average value of $\Delta\omega(t)$ and $\beta(\tau)$ is the phase of the oscillation.

The frequency shift of the oscillation is given by $\Delta = \langle \Delta\omega \rangle$. The line width $\tilde{\gamma}_{RS}$ can be calculated if the function $\langle e^{i\psi(\tau)} \rangle$ is known.

We discuss now the main properties of the distribution (11). If τ_0 is the average value of the interval τ_i during which $\Delta\omega(t)$ retains its sign, then τ_0 can be defined as the correlation time of a random function $\Delta\omega(t)$, i.e., when $\tau \gg \tau_0$ we have

$$\overline{\Delta\omega(t)\,\Delta\omega(t+\tau)} \simeq 0.$$

We consider the case of very fast modulation, when the change $\Delta\beta_i$ of the oscillation phase during the time τ_i is small:

$$\langle \Delta\beta_i \rangle \simeq \langle \Delta\omega \rangle \tau_0 \ll 1. \tag{15}$$

It is shown in [80] that the line width $\tilde{\gamma}_{RS}$ can be defined under the condition (15) as

$$\tilde{\gamma}_{RS} = {}^1\!/_4\, \langle (\Delta\omega - \langle \Delta\omega \rangle)^2 \rangle \tau_0. \tag{16}$$

Let us prove this. In the calculation of $\tilde{\gamma}_{RS}$ we choose a time interval τ [see (11) and (14)] for which the corresponding phase shift is $\beta(\tau) \sim 1$, i.e., $\tau \gg \tau_0$. For such τ we have

$$\beta(\tau) \approx \sum_{i=1}^N \Delta\beta_i, \quad N \simeq \frac{\tau}{\tau_0} \gg 1;$$
$$\langle \Delta\beta_i \Delta\beta_j \rangle = \delta_{ij} \langle \Delta\beta^2 \rangle. \tag{17}$$

This means that the distribution function $P(\beta, \tau)$ is Gaussian

$$P(\beta, \tau) = \frac{1}{\sqrt{2\pi \langle \beta^2(\tau) \rangle}}\, e^{-\frac{\beta^2}{2\langle \beta^2(\tau) \rangle}}. \tag{18}$$

Therefore

$$\langle e^{i\beta(\tau)} \rangle = \int e^{i\beta} P(\beta, \tau)\, d\beta = e^{-\frac{1}{2}\langle \beta^2(\tau) \rangle}, \tag{19}$$

$$\langle \beta^2(\tau) \rangle = \sum_{i,j}^N \langle \Delta\beta_i \Delta\beta_j \rangle \simeq N \langle \Delta\beta^2 \rangle \simeq \frac{\tau}{\tau_0} \langle \Delta\beta^2 \rangle, \tag{20}$$

$$\langle \Delta\beta^2 \rangle \simeq \langle (\Delta\omega - \langle \Delta\omega \rangle)^2 \rangle\, \tau_0. \tag{21}$$

The final equation for $\langle e^{i\beta(\tau)} \rangle$ is

$$\langle e^{i\beta(\tau)} \rangle = e^{-1/2\, \langle (\Delta\omega - \langle \Delta\omega \rangle)^2 \rangle \tau_0 \cdot \tau}. \tag{22}$$

Equations (11), (14), and (22), as shown in [80], yield a Lorentz distribution defined by Eq. (16) and having a width $\tilde{\gamma}_{RS}$.

If the phase randomization of the molecular oscillation in the liquid is due to motion and collisions of the molecules, then we can determine the temperature dependence of τ_0 in the liquid by using the diffusion formula

$$\tau_0 \sim \frac{a^2}{D} \sim \eta\, \frac{a^3}{kT}, \tag{23}$$

where D is the diffusion coefficient.

The line width is then

$$\tilde{\gamma}_{RS} \sim \langle (\Delta\omega - \langle \Delta\omega \rangle)^2 \rangle\, \eta\, \frac{a^3}{kT}. \tag{24}$$

It is difficult to say anything definite concerning the temperature dependence of $\langle (\Delta\omega - \langle \Delta\omega \rangle)^2 \rangle$, but there are no reasons for this quantity to increase rapidly with temperature. It is most probable that the temperature dependence of $\tilde{\gamma}_{RS}$ is determined by the factor $\eta a^3/kT$. This quantity decreases with temperature in liquids.

In carbon disulfide, in accordance with the results of Shapiro and Broida [81], the value of τ_0 varies in the temperature range 300 to 161°K from $0.4 \cdot 10^{-12}$ to $8 \cdot 10^{-12}$ sec. If we assume that

$$\langle(\Delta\omega - \langle\Delta\omega\rangle)^2\rangle \simeq \langle\Delta\omega\rangle^2, \tag{25}$$

then it is quite reasonable to expect a narrowing of the line by 0.3 cm^{-1} when the temperature rises from 161 to 300°K. Of course, the narrowing can be observed if it is not cancelled by the increase of the line width on account of the reorientation of the molecules. This condition is satisfied in carbon disulfide.

The similarity of the temperature dependences of the RS and SRS line widths in carbon disulfide and in calcite, and the existence of the minimum at one and the same temperature in carbon disulfide, gives grounds for assuming that the processes that determine the course of the temperature dependence of the RS line width determine the course of the temperature dependence of the summary SRS line width.

The difference between the summary spectral width of the first Stokes component of SRS forward and backward, which was observed by us experimentally, can be due to the difference between the SRS intensities propagating in these directions. The question of the asymmetry of the SRS in these directions was discussed in detail by Kudryavtseva [82].

We have established that for all the investigated substances, regardless of the value of the Kerr constant, if the SRS is observed with high-resolution instrument, the contour of the SRS line consists of individual components. The width of the fine-structure component is on the order of the instrumental-function width and is smaller by two orders of magnitude than the RS line width. It does not depend on the excitation energy or on the temperature of the substance. The distances between the fine-structure components do not correspond to distances between the SMBS components or to the distances between the ruby-laser modes.

The existence of fine-structure components is not the consequence of the selective action of the different details of the experimental setup. Favoring this assumption is the reproducibility of the results at different geometry of the experiment. When we use a calcite crystal with parallel end faces we observe no substantial changes in the spectral distribution, although the threshold of the SRS decreased in this case compared with the threshold of a sample having end faces tilted 10°.

The characteristic features of the fine structure observed by others are, on the one hand, certain randomness in the onset of the number and the alternation of the intensities in space, and on the other the quasiperiodicity of their arrangement. The number of components and the average value of the distance between them remain the same in 70% of the cases. If a scatterer is placed in front of the etalon, then the structure averages out and the line contour becomes smooth.

The fine structure observed in the spectral distribution of the intensity of the SRS components, which was investigated in detail for the first time ever in the present paper, is apparently the result of the enhancement of the fluctuations of the spontaneous noise of ordinary RS. This is attested by the already indicated randomness of the observed distributions, the very narrow width of the individual components (on the order of 10^{-2}–10^{-3} cm^{-1}), and the nondependence of the width of the component on the experimental conditions. The gain contour is the ordinary RS line. It should be noted that if the structure observed by us constitutes simply amplified ordinary RS noise bursts, then the spectral width of the individual noise component should not depend on the temperature of the scattering medium. In this case the spectral width γ_{SRS} of an individual fine-structure component is connected with the duration τ_p of the pulse of the first Stokes component by the relation

$$\gamma_{SRS}\tau_p = \text{const.}$$

It was therefore of interest to trace simultaneously the variation of the spectral width of an individual structure component and of the duration of the pulse of the first SRS Stokes component when the temperature of the medium changes. We have performed such measurements on the first SRS Stokes component of calcite, whose temperature was varied from 77 to 380°K. The measurements were made for SRS light scattered in the forward direction. Figure 25 shows the results of the measurements of the duration of the pulse of the first SRS Stokes component at different temperatures. It is seen from the figure that the pulse duration increases only slightly from 12.5 to 17.5 nsec. In the entire temperature interval we observed in the first Stokes component a fine spectral structure whose component widths remain practically unchanged at 0.01-0.02 cm^{-1}. The results of these measurements do not contradict the premise that the fine-structure components are due to noise.

In a recent papers, Ishchenko and co-authors [83] described the results of experimental investigations of the superradiance line splitting of pulsed N_2 and N_2^+ lasers, as well as Tl and Ne lasers [84, 85]. An analysis

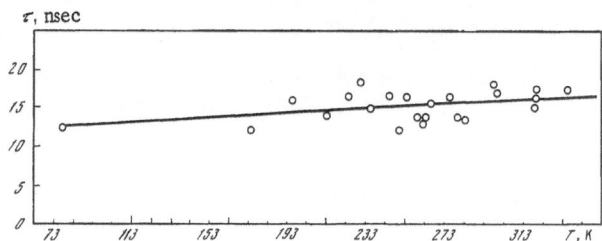

Fig. 25. Dependence of the duration of the pulse of
the first SRS Stokes component of calcite in the forward direction on the temperature of the medium.

of the results of [83-85], as well as the results for SRS obtained in [79], has enabled the authors of [83] to conclude that such a splitting of the lines is a property of all pulsed sources, regardless of the type of radiating
system and of the excitation method. Thus, the observed phenomenon is quite universal and is due to the amplification of the fluctuation bursts of the spontaneous emission. The resultant periodicity of the spectral distribution of the intensity seems to indicate an obviously nonlinear spectrum transformation. Several possible
physical mechanisms of the periodization of the fluctuation structure of the spectrum are indicated in [83]
(splitting of the gain contour in the presence of a strong field, four-photon resonant scattering of light, photon
echo). Each of these factors can lead to periodicity with random period.

Our experimental investigations have shown that in the case of SRS of light there are produced in the
medium large power densities of the scattered radiation, thus creating favorable conditions for the development of various kinds of nonlinear processes, particularly self-focusing of the light. If the propagation of the
light in the medium is in the form of a multifocus structure, then we cannot exclude the possibility that the
interference of the radiation from two or more focal regions can lead to a quasiperiodic structure in the summary frequency distribution of the intensity.

The temperature dependence of the period of the fine structure in calcite can be due to the change in the
regime of the self-focusing (change in the dimensions of the self-focusing dots and in the distances between
them).

There are also several physical mechanisms that can result in a quasiperiodic structure of the same
scale in the spectral distribution of the SRS. Brewer [4] believes that the appearance of a structure in the SRS
spectrum can be due to self-focusing of light, since self-focusing in a medium produces conditions for stimulated 2- and 4-photon Rayleigh-wing scattering [33], as a result of which additional frequency components appear in the spectrum.

The periodic structure observed in [86] in the spectrum of the scattered radiation is the result of Stark
modulation of the molecular vibrations (self-focusing is not considered there). It is shown in that paper that
with increasing excitation energy the number of components in the spectrum increases. The authors of [86]
expect the Stark modulation to manifest itself in the narrow scattering lines in liquid nitrogen and oxygen.

It is shown in [87] that the growth of the intensity of the scattered radiation is nonmonotonic and has an
oscillatory character (in the given pump-intensity approximation). The appearance of oscillations in the spectrum has a threshold. The authors of [87] assume that this theory describes well the increase of the number of
components in the spectral distribution of the intensity of the first SRS Stokes component of liquid nitrogen,
which was observed by us in experiment [79].

The theoretical papers listed above do not explain many singularities of the spectral distribution of the
intensity of the SRS components (the small width of the component, the stochasticity of the distribution of the
intensity in space and over the components, etc.). The totality of the experimental results of the present paper
shows that the most probable cause of the fine structure in the SRS spectra is the nonlinear conversion of the
spontaneous RS noise in the scattering medium although the physical mechanism that leads to the appearance of
regularity in the arrangement of the components is at present not quite clear.

The experimental results connected with the development of self-focusing of the SRS components and
the appearance of large spectral broadenings can be fitted within the framework of a theory that describes
the propagation of light in a medium in the form of a multifocus structure [12]. In liquid nitrogen and calcite,
in nitrobenzene, and in carbon disulfide, under excitation conditions close to the threshold, the development

of self-focusing of SRS of light takes place in a quasistationary* regime. The picture of the development of self-focusing, as stated in Section 1 of Chapter V, manifests itself most clearly in the SRS components in carbon disulfide. In the cross sections of the beams of the first SRS Stokes component near the exit face of the sample there are observed self-focusing dots, while rings are observed in the cross sections of the beams of the higher Stokes and anti-Stokes components. The fact that at an excitation power approximately from 0.5 to 2.5 MW there appear in the cross sections of the beams of the higher stokes and anti-Stokes components of the SRS of carbon disulfide a large number of rings of various diameters, corresponding to a section of cones with identical aperture angle, is evidence that the cones are radiated from definite points of space that do not lie in the same plane, but belong to one type of radiation. At excitation powers close to the threshold, an agreement is observed between the position of the centers of the rings of the higher Stokes and anti-Stokes SRS dots, on the one hand, and the self-focusing dots of the first Stokes component, on the other. This indicates that the cones are radiated from the self-focusing regions. The appearance of systems of concentric rings indicates that the self-focusing regions that are produced by the radiation cones can be located on a single axis.

The vertices of the cones of the emission of the SRS components of higher order can be the turning points of the foci of the multifocus structure produced in the beam of the first SRS Stokes component when the SRS is nonstationary. With increasing excitation energy, a decrease is observed in the diameters of the rings in the sections of the SRS-component beams, until autonomous focal regions of the corresponding component appear in the investigated section of the cell. This decrease of the diameters can be attributed to the approach of the turning points of the multifocus structure of the first SRS Stokes component to the exit face of the cell. When a definite threshold excitation power is reached, large spectral broadenings of the SRS components appear and indicate a transition of the SRS into a nonstationary regime, as well as the appearance of ultrashort pulses [43] and of an autonomous multifocus structure in them. An estimate of the pulse durations from the experimentally observed broadenings yields values on the order of 10^{-11}-10^{-13} sec. The characteristic pictures of the asymmetry of the broadening, and in a number of cases the quasiperiodic structure in them, can be explained, according to [12], as being due to the different velocity of the focal regions and to interference of the radiation emerging from different focal points.

Large spectral broadenings on the order of 100 cm^{-1} appear in points with minimal diameter on the order of 10 μm in carbon disulfide and nitrobenzene, and do not appear in calcite and in liquid nitrogen, in which self-focusing points of the same diameter are observed. This may be due to the different values of the radiation-power density contained in the focal region. According to measurements of Kudryavtseva [82], the power density at the self-focusing points in carbon disulfide and in nitrobenzene is larger by one order of magnitude than in liquid nitrogen and in calcite. A nonstationary SRS regime in liquid nitrogen and calcite is not reached under any experimental conditions.

The emission of the higher SRS Stokes and anti-Stokes components observed in the case of saturation in calcite and in liquid nitrogen is well described by the theoretical relations derived by Lugovoi [19]:

$$k_m + m k_{-1}^{(m)} = (m + 1) k_0,$$
$$k_{-m} + (m - 1) k_0 = m k_{-1},$$

(26)

where k_m are the wave vectors.

The radiation described by these relations is called radiation of class I, while the radiation cones that do not satisfy these relations are called radiations of class II.

The complicated picture of the angular distribution of the SRS components of higher orders, which is most clearly observed in carbon disulfide, has long attracted the attention of the investigators and did not find a complete theoretical explanation. The existing assumption that radiation of this type comes from self-focusing filaments was not directly confirmed by experiment. In the present paper, in Section 1 of Chapter V, it is shown experimentally that the emission of the higher SRS Stokes and anti-Stokes components comes from regions of self-focusing of the first SRS Stokes components, when the SRS is quasistationary. The origin of the

---

*The characteristic time of establishment of the stationary SRS τ_{SRS} can be estimated from the condition $\tau_{SRS} \sim k\tau_{RS}$, where k is the gain, and τ_{RS} is the reciprocal of the half-width of the line of ordinary RS and is on the order of 10^{-10} sec for carbon disulfide. The characteristic time of stay of the focus in a given point of the medium (τ) is approximately 10^{-10} sec [12]. Therefore to excite SRS near the threshold the gain over the length of the focal region is $e^{0.01}$. Then $\tau_{SRS} = 10^{-12}$ sec and $\tau > \tau_{SRS}$, meaning that the process is quasistationary. If the SRS saturates, the gain is e^{100}, and the settling time is $\tau_{SRS} = 10^{-8}$ sec, so that under the conditions considered $\tau < \tau_{SRS}$, and the SRS process is nonstationary.

emission cones produced in carbon disulfide can be understood on the basis of concepts developed by Lugovoi and Prokhorov [88]. Depending on the exciting-radiation power, various self-focusing regimes accompany the SRS of light. Changes occur in the dimensions, in the velocities of the focal regions, and in the distances between them. In the case of a strongly elongated caustic of the radiation in the cell with the medium (waveguide filament), radiation of the Cerenkov type is produced [89, 90]. The angle of the radiation cone is in this case for the first anti-Stokes component

$$\theta_{Cer} = \sqrt{\frac{2\sqrt{k_i}}{k_1}}.$$

The angles of emission of class II can be interpreted as radiation from moving focal regions in the presence of a multifocus structure in the first Stokes component, and lie in the range

$$\theta_{cl.I} < \theta_{cl.II} < \theta_{Cer}. \tag{27}$$

In carbon disulfide, the emission angles of the anti-Stokes components of class I are equal to [91]

$$\theta_{1cl.I} = 0.0235 \text{ rad}, \qquad \theta_{2cl.I} = 0.0464 \text{ rad}.$$

The angles of the corresponding cones of the Cerenkov type are

$$\theta_{1Cer} = 0.037 \text{ rad}, \qquad \theta_{2Cer} = 0.066 \text{ rad}.$$

The experimentally observable emission angles are

$$\theta_{1cl.II} = 0.032 \text{ rad},$$
$$\theta_{2cl.II} = 0.0432 \text{ rad}, \ 0.061 \text{ rad},$$

and lie in the region described by the conditions (27).

If a multifocus structure exists in the exciting radiation then, according to [88], the emission cones for the first anti-Stokes component should lie between the limits

$$\theta_{Cer} < \theta < 2\sqrt{\frac{k_i \sqrt{k_0}}{k_1 k_1}} = \varphi. \tag{28}$$

So far, such values of the angle were not observed in experiment. The authors of [88] believe that the absence of radiation at angles (28) is apparently connected with the fact that the multifocus structure in the exciting radiation occurs under essentially nonstationary conditions (relative to SRS) [12].

As shown by our experimental results, at high excitation energies the SRS becomes nonstationary and large spectral broadenings appear. Under these conditions the experimentally observed emission angles of the SRS anti-Stokes components are located in the range (28).

First anti-Stokes component $\theta_{Cer}(0.037) < 0.0468 < \varphi\ (0.05)$.

Second anti-Stokes component $\theta_{Cer}(0.066) < 0.0713 < \varphi\ (0.095)$.

Consequently, the theoretical estimates of the angles of the anti-Stokes components, obtained on the premise that the light propagates in a nonlinear medium in the form of focal regions, are close to the values obtained for the first and second SRS anti-Stokes components of carbon disulfide.

CONCLUSIONS

In the present experimental paper we report the first systematic investigations of the spectral distribution of the intensity of SRS components of a number of substances with different values of the Kerr constant. We investigated the influence exerted on the spectral distribution of the SRS by the thickness of the scattering layer, by the energy of the exciting radiation, and by the temperature of the medium. It is shown on the basis of extensive experimental material that the spectral width of the first SRS Stokes component is not described by the theoretical relation obtained from the premise that the emission line becomes narrower when the radiation passes through a nonlinear medium. Near the threshold, the spectral width of the first Stokes component is lower by one order of magnitude than the width obtained theoretically, and is smaller by two orders of magnitude than the width of the RS line. When the intensity of the exciting radiation is increased, additional components appear in the spectral distribution, and the width of the contour of the envelope of the fine-structure components increases and approaches the theoretical value. The spectral width of an individual fine-structure component remains constant.

We performed for the first time ever detailed investigations of the fine structure observed in the spectral distribution of the intensity of the SRS components. A number of experiments were performed that made it possible to conclude that the cause of the structure may be spontaneous noise. The possible mechanisms of the onset of a periodic fine structure, which are connected with nonlinear conversion of the noise spectrum, are considered.

We investigated the temperature dependence of the width of the summary spectral distribution of the intensity of the SRS components. On the basis of the result it is concluded that the temperature dependence of the width of the summary spectral distribution of the intensity of the SRS components is determined by mechanisms that lead to broadening of the lines of the ordinary SR (reorientational motion of the molecules, randomization of the phase of the molecular oscillations).

Systematic experimental investigations were performed, for the first time ever, of the connection between self-focusing of light and the spectral and angular characteristics of the SRS. It is shown that the onset of large spectral broadenings at the SRS self-focusing points and the angular directivity of the SRS components in media with large Kerr constants can be explained from the point of view of the premise that the high-power optical radiation propagates in the medium in the form of moving focal regions.

In conclusion I am grateful to A. I. Sokolovskaya and M. M. Sushchinskii for guidance and constant interest in the work.

LITERATURE CITED

1. T. K. Gustafson, J. P. Taran, H. A. Haus, J. R. Lifsits, and P. L. Kelley, Phys. Rev., 177:306 (1969).
2. A. C. Cheung, D. M. Rank, R. Y. Chiao, and C. H. Townes, Phys. Rev. Lett., 20:786 (1968).
3. F. Shimizu, Phys. Rev. Lett., 19:1097 (1967).
4. R. G. Brewer, Phys. Rev. Lett., 19:8 (1967).
5. H. P. H. Grieneisen and C. A. Sacchi, Bull. Am. Phys. Soc., 12:686 (1967).
6. R. G. Brewer, J. R. Lifsits, E. Garmire, R. Y. Chiao, and C. H. Townes, Phys. Rev., 166:326 (1968).
7. R. Cubeddu, R. Polloni, C. A. Sacchi, and O. Svelto, Phys. Rev. A, 2:1955 (1970); R. Cubeddu and F. Zaraga, Opt. Commun., 3:310 (1971).
8. F. De Matrini, C. H. Townes, T. K. Gustafson, and P. L. Kelley, Phys. Rev., 164:312 (1967).
9. N. Bloembergen and P. Lallemand, Phys. Rev. Lett., 16:81 (1966).
10. P. Lallemand, Appl. Phys. Lett., 8:276 (1966).
11. N. Bloembergen, P. Lallemand, and A. Pine, IEEE J. Quantum Electron. QE-2:246 (1966).
12. V. N. Lugovoi and A. M. Prokhorov, Usp. Fiz. Nauk, 111:203 (1973).
13. V. A. Chirkov, V. S. Gorelik, G. V. Peregudov, and M. M. Sushchinskii, Pis'ma Zh. Éksp. Teor. Fiz., 10:416 (1969).
14. G. Eckhardt, R. W. Helwarth, F. J. McClung, S. E. Schwarz, D. Weiner, and E. J. Woodbury, Phys. Rev. Lett., 9:455 (1962); Electron. Des., 11:28 (1963).
15. M. Geller, D. P. Bortfeld, and W. R. Sooy, Appl. Phys. Lett., 3:36 (1963).
16. G. Bret, Ann. Radioelectron., 22:236 (1967).
17. A. D. Kudryavtseva, A. I. Sokolovskaya, and M. M. Sushchinskii, Zh. Éksp. Teor. Fiz., 59:1556 (1970).
18. V. A. Zubov, P. P. Kurcheva, and M. M. Sushchinskii, Kratk. Soobshch. Fiz., No. 1:45 (1971).
19. V. N. Lugovoi, Introduction to the Theory of Stimulated Raman Scattering [in Russian], Nauka, Moscow (1968).
20. M. M. Sushchinskii, Raman Spectra of Molecules and Crystals [in Russian], Nauka, Moscow (1969).
21. N. Bloembergen, Am. J. Phys., 35:989 (1967).
22. Yu. E. D'yakov, Pis'ma Zh. Éksp. Teor. Fiz., 10:545 (1969).
23. R. Hellwarth, Phys. Rev., 130:1850 (1963); Curr. Sci. (India), 3:129 (1964).
24. N. Bloembergen and Y. R. Shen, Phys. Rev., 133:A37 (1964; 137:A1787 (1965).
25. J. B. Grun, A. K. McQuillan, and B. P. Stoicheff, Phys. Rev., 180:61 (1969).
26. D. von der Linde, M. Maier, and W. Kaiser, Phys. Rev., 178:11 (1969).
27. S. A. Akhmanov, A. P. Sukhorukov, and R. V. Khokhlov, Usp. Fiz. Nauk, 93:19 (1967); Zh. Éksp. Teor. Fiz., 50:1537 (1966); Zh. Éksp. Teor. Fiz., 51:296 (1966).
28. V. I. Talanov, Izv. Vyssh. Uchebn. Zaved., Radiofiz., 7:564 (1964); Pis'ma Zh. Éksp. Teor. Fiz., 11:303 (1970).
29. P. L. Kelley, Phys. Rev. Lett., 15:1005 (1965).

30. V. I. Bespalov and V. I. Talanov, Pis'ma Zh. Éksp. Teor. Fiz., 3:471 (1966).
31. Ya. B. Zel'dovich and Yu. P. Raizer, Pis'ma Zh. Éksp. Teor. Fiz., 3:137 (1966).
32. N. F. Pilipetskii and A. R. Rustamov, Pis'ma Zh. Éksp. Teor. Fiz., 2:88 (1965).
33. D. I. Mash, V. V. Morozov, V. S. Starunov, and I. L. Fabelinskii, Pis'ma Zh. Éksp. Teor. Fiz., 2:41 (1965).
34. B. P. Stoicheff, Phys. Lett., 7:186 (1963).
35. W. J. Jones and B. P. Stoicheff, Phys. Rev. Lett., 13:657 (1964).
36. A. L. Dyshko, V. N. Lugovoi, and A. M. Prokhorov, Pis'ma Zh. Éksp. Teor. Fiz., 6:655 (1967).
37. V. N. Lugovoi and A. M. Prokhorov, Pis'ma Zh. Éksp. Teor. Fiz., 7:153 (1968).
38. V. V. Korobkin and R. V. Serov, Pis'ma Zh. Éksp. Teor. Fiz., 6:642 (1967).
39. V. E. Zakharov, V. V. Sobolev, and V. S. Synakh, Pis'ma Zh. Éksp. Teor. Fiz., 14:564 (1971).
40. V. N. Lugovoi and A. M. Prokhorov, Pis'ma Zh. Éksp. Teor. Fiz., 12:478 (1970).
41. G. K. L. Wong and Y. R. Shen, Appl. Phys. Lett., 21:162 (1972).
42. V. V. Korobkin, V. N. Lugovoi, A. M. Prokhorov, and R. V. Serov, Pis'ma Zh. Éksp. Teor. Fiz., 16:595 (1972).
43. M. Maier, W. Kaiser, and J. A. Giordmaine, Phys. Rev. Lett., 17:1275 (1966).
44. M. M. T. Loy and Y. R. Shen, Appl. Phys. Lett., 19:285 (1971).
45. V. V. Korobkin, A. M. Prokhorov, R. V. Serov, K. F. Shipilov, and T. A. Shmanov, Phys. Lett., 47A:381 (1974).
46. A. I. Sokolovskaya, A. D. Kudryavtseva, T. P. Zhbanova, and M. M. Sushchinskii, Zh. Éksp. Teor. Fiz., 53:429 (1967).
47. A. D. Kudryavtseva, A. I. Sokolovskaya, and M. M. Sushchinskii, Kvantovaya Élektron. (Moscow), No. 7, 73 (1972).
48. A. I. Sokolovskaya, E. A. Morozova, A. D. Kudryavtseva, and M. M. Sushchinskii, Kvantovaya Élektron. (Moscow), No. 4, 76 (1973).
49. A. D. Kudryavtseva and A. I. Sokolovskaya, Kvantovaya Élektron. (Moscow), 1:964 (1974).
50. N. V. Zubova, M. M. Sushchinskii, and V. A. Zubov, Pis'ma Zh. Éksp. Teor. Fiz., 2:63 (1965).
51. N. V. Zubova, N. P. Kuz'mina, V. A. Zubov, M. M. Sushchinskii, and I. K. Shuvalov, Zh. Éksp. Teor. Fiz., 51:101 (1966).
52. I. I. Kondilenko, P. A. Korotkov, and V. I. Maly, Phys. Lett., A42:72 (1972).
53. G. A. Askar'yan, Pis'ma Zh. Éksp. Teor. Fiz., 4:400 (1966).
54. V. A. Chirkov, G. V. Peregudov, and M. M. Sushchinskii, Abstracts of the 17th All-Union Congress of Spectroscopy [in Russian], Izd. Inst. Fiz. AN BSSR, Minsk (1971), Part 2, p. 49.
55. G. S. Landsberg, P. A. Bazhulin, and M. M. Sushchinskii, Principal Parameters of Raman Spectra of Hydrocarbons [in Russian], Izd. AN SSSR, Moscow (1956).
56. S. E. Frish, Spectroscopy Techniques [in Russian], Izd. LGU, Leningrad (1936).
57. S. Tolansky, High-Resolution Spectroscopy, Methuen, London (1947).
58. S. G. Rautian, Usp. Fiz., Nauk, 66:475 (1958).
59. P. A. Bazhulin, S. G. Rautian, A. I. Sokolovskaya, and M. M. Sushchinskii, Zh. Éksp. Teor. Fiz., 29:822 (1955).
60. E. K. Kazakova, A. V. Kraiskii, V. A. Zubov, M. M. Sushchinskii, and I. K. Shuvalov, Kratk. Soobshch. Fiz., No. 7, 42 (1970).
61. V. S. Gorelik and M. M. Sushchinskii, Fiz. Tverd. Tela (Leningrad), 12:1475 (1970).
62. Y. Kato and H. Takuma, J. Opt. Soc. Am., 61:347 (1971).
63. F. J. McClung and D. Weiner, J. Opt. Soc. Am., 54:641 (1964).
64. G. L. Brekhovskikh, A. I. Sokolovskaya, and V. A. Seleznev, Zh. Prikl. Spektrosk., 19:44 (1973).
65. A. D. Kudryavtseva, E. A. Morozova, and M. M. Moiseenko, Kratk. Soobshch. Fiz., No. 10, 31 (1973).
66. G. L. Brekhovskikh, Kratk. Soobshch. Fiz., No. 11, 23 (1974).
67. W. R. L. Clements and B. P. Stoicheff, Appl. Phys. Lett., 12:246 (1968).
68. K. Park, Phys. Lett., 22:39 (1966).
69. R. W. Hellwarth, Appl. Opt., 2:847 (1963).
70. G. Eckhardt, IEEE J. Quantum Electron., QE-2:1 (1966).
71. I. I. Sobel'man, Tr. Fiz. Inst. Akad. Nauk SSSR, 9:315 (1958).
72. Kh. E. Sterin, Tr. Fiz. Inst. Akad. Nauk SSSR, 9:15 (1958).
73. A. I. Sokolovskaya, Tr. Fiz. Inst. Akad. Nauk SSSR, 27:63 (1964).
74. A. V. Rakov, Tr. Fiz. Inst. Akad. Nauk SSSR, 27:111 (1964).
75. N. I. Rezaev, Materials of the 10th All-Union Conference on Spectroscopy [in Russian], Vol. 1, Izd. L'vov. Gos. Univ. (1957), p. 230; Candidate's Dissertation, Moscow State Univ. (1958).

76. G. V. Mikhailov, Materials of the 10th All-Union Conference on Spectroscopy [in Russian], Vol. 1, Izd. L'vov. Gos. Univ. (1957), p. 227.

77. B. P. Stoicheff, Can. J. Phys., 1976.

78. A. I. Sokolovskaya, E. A. Morozova, and A. D. Kudryavtseva, Zh. Prikl. Spektrosk., 18:122 (1973).

79. E. A. Morozova, A. I. Sokolovskaya, and M. M. Sushchinskii, Zh. Éksp. Teor. Fiz., 65:2161 (1973).

80. V. A. Alekseyev and I. I. Sobelman, Can. J. Phys., 1976.

81. S. L. Shapiro and H. P. Broida, Phys. Rev., 154:129 (1967).

82. A. D. Kudryavtseva, Tr. Fiz. Inst. Akad. Nauk SSSR, 99:49 (1977).

83. V. I. Ishchenko, V. N. Lisitsyn, A. M. Razhev, S. G. Rautian, and A. M. Shalagin, Pis'ma Zh. Éksp. Teor. Fiz., 19:669 (1974).

84. F. A. Korolev, G. V. Abrosimov, and A. I. Odintsov, Opt. Spektrosk., 33:725 (1972).

85. G. V. Abrosimov, N. G. Andreev, and A. I. Odintsov, Vestn. Mosk. Gos. Univ., Fiz., Astron., 14:287 (1973).

86. S. K. Potapov, B. A. Medvedev, M. A. Kovner, and I. L. Klyukach, Kvantovaya Élektron. (Moscow), No. 2 (14), 416 (1973).

87. V. I. Emel'yanov and Yu. D. Klimontovich, Zh. Éksp. Teor. Fiz., 68:929 (1975).

88. V. I. Lugovoi and A. M. Prokhorov, Zh. Éksp. Teor. Fiz., 69:84 (1975).

89. A. Szöke, Bull. Am. Phys. Soc., 9:490 (1964).

90. V. N. Lugovoi and I. I. Sobel'man, Zh. Éksp. Teor. Fiz., 58:1283 (1970).

91. E. Garmire, Phys. Lett., 17:251 (1965).

STATIONARY SRS REGIMES IN THE FIELDS OF ULTRASHORT PULSES

T. M. Makhviladze and M. E. Sarychev

It is established that in ultrashort pulsed strong fields there are produced unique stationary SRS regimes that are not accompanied by amplification (SRS solitons). It is shown that, depending on the dispersion properties of the medium and on the initial conditions, different soliton regimes are realized — both single pulses and periodic trains. The influence of the relaxation processes on the propagation of SRS solitons and the stability of the soliton regimes to the action of small perturbations of various types are investigated. The results are applicable to the case of two-photon resonant absorption of pulses of different frequencies.

1. INTRODUCTION

The effect of self-induced transparency, which occurs when powerful ultrashort pulses of light pass through a medium, is extensively investigated of late [1]. The interest in this phenomenon is due both to the fact that it can be used to extract additional information on the optical characteristics of substances (determination of the radiative-transition constants and of the times of transverse and longitudinal relaxation), and with the possibility of obtaining ultrashort pulses. The theory of the effect of coherent transparentization of a medium with single-photon resonant absorption was first presented by McCall and Hahn [2]. It is shown in [1, 3, 4] that this effect takes place also in media with two-photon resonant and nonresonant absorption.

In the present paper we consider the effect of self-induced transparency in stimulated Raman scattering (SRS) under conditions of interaction of ultrashort pulses of the exciting Stokes radiation in a lossless medium (SRS solitons). This effect arises only when the duration of both pulses is much shorter than the transverse-relaxation time T_2. Its reason is that the energy absorbed by the medium from the pulses is then coherently returned to the fields on account of stimulated scattering, making possible a stationary scattering regime wherein the shapes and amplitudes of both pulses remain unchanged.

As will be shown below, the effect has a threshold and occurs only in sufficiently strong fields. At intensities above threshold, a large change takes place in the level populations in the course of scattering, and this distinguishes greatly the situation from SRS in the field of quasistatic pulse or ultrashort pulses with below-threshold intensity (see [5]). However, even if the threshold conditions are satisfied on entering the medium, in order to ascertain where the SRS pulses will change into solitons it is necessary to solve the complete system of nonstationary equations that describe their temporal evolution. This can be done only in the particular case of "proportional" input pulses. Nor can a final answer to the question be obtained by numerically solving the nonstationary equations. It is therefore particularly important to determine the stability of the motion of SRS solitons to small perturbations.

In the present paper we find all the possible soliton regimes that can exist in the case of combined interaction of ultrashort pulses in a medium without losses (Sections 2 and 3). Realization of a particular type of soliton depends on the initial conditions and on the dispersion properties of the medium. In Section 3 we consider the influence of the relaxation processes on the propagation of SRS solitons, while in Section 4 we investigate the stability of certain soliton regimes to various types of small perturbations. The exposition is based on earlier works by the authors [6].

2. FUNDAMENTAL EQUATIONS

We assume that the scattering is from one pair of levels of the scattering molecules, and consider for simplicity a one-dimensional problem. In the envelope approximation, the system of equations describing the evolution of the pulses of the exciting and first Stokes SRS radiation consists of the abbreviated Maxwell's equations for the field amplitudes and of the equations of motion for the average values of the polarization. In the case of resonance (ω_i, ω_S are the frequencies of the exciting and Stokes fields, ω_V is the frequency of the working transition), namely

$$\omega_i - \omega_s = \omega_v ,$$

(1)

this system takes the form* [6]

$$\frac{\partial \mathscr{E}_s}{\partial z} + \frac{1}{c_s} \frac{\partial \mathscr{E}_s}{\partial t} = -\frac{\lambda \mu_0 c_s \omega_s N_V}{2} v \mathscr{E}_i, \quad \frac{\partial \mathscr{E}_i}{\partial z} + \frac{1}{c_i} \frac{\partial \mathscr{E}_i}{\partial t} = \frac{\lambda \mu_0 c_i \omega_i N_V}{2} v \mathscr{E}_s,$$

$$\frac{du}{dt} = -\frac{u}{T_2}, \quad \frac{dv}{dt} = -\frac{v}{T_2} + \frac{\lambda}{\hbar} \mathscr{E}_i \mathscr{E}_s W, \quad \frac{dW}{dt} = -\frac{\lambda}{\hbar} \mathscr{E}_i \mathscr{E}_s v - \frac{W - W^{eq}}{T_1},$$

(2)

where $\mathscr{E}_i$, $\mathscr{E}_s$ are the real amplitudes of the fields of the exciting and Stokes radiation; λ is the matrix element of the scattering; u and v are the amplitudes of the transverse polarization; W is the half-difference of the populations of the upper and lower levels; W^{eq} is the equilibrium value of W; T_1 and T_2 are the times of longitudinal and transverse relaxation; $\mu_0 = 4\pi/c^2$; N_V is the density of the scattering molecules; $c_{i,s} = (\partial \omega / \partial k)_{i,s}$ are the group velocities of the pump and Stokes waves (k is the modulus of the wave vector). To consider ultrashort pulses with durations much less than T_2 and T_1 we assume, besides (1), that $T_1 = T_2 = \infty$ (allowance for finite T_1 and T_2 will be made below). The system (2) can then be easily reduced to the form

$$\frac{\partial \mathscr{E}_s}{\partial z} + \frac{1}{c_s} \frac{\partial \mathscr{E}_s}{\partial t} = -\varkappa_s \mathscr{E}_i \sin \varphi, \quad \frac{\partial \mathscr{E}_i}{\partial z} + \frac{1}{c_i} \frac{\partial \mathscr{E}_i}{\partial t} = \varkappa_i \mathscr{E}_s \sin \varphi,$$

$$\varphi = \frac{\lambda}{\hbar} \int \mathscr{E}_i(z, t') \mathscr{E}_s(z, t') \, dt', \quad \varkappa_i = \frac{1}{2} \lambda \mu_0 c_i N_V W^{eq} \omega_i;$$

$$\varkappa_s = \frac{1}{2} \lambda \mu_0 c_s N_V W^{eq} \omega_s$$

(3)

The quantity $\varphi(z, t)$ has the meaning of the angle of rotation of the polarization vector.

The general soliton solution of the system (3) is of the form

$$\varphi = \varphi_0 \left(t - \frac{z}{V} \right), \quad \mathscr{E}_s = \mathscr{E}_s \left(t - \frac{z}{V} \right), \quad \mathscr{E}_i = \mathscr{E}_{0i} \left(t - \frac{z}{V} \right),$$

(4)

where V is the group velocity of the pulses and is the same for $\mathscr{E}_i$ and $\mathscr{E}_s$. It is easily seen that only equality of the group velocities ensures the possible existence of pulses whose shape and intensity do not change; in the opposite case the coherence of their interaction is violated even over a length on the order of the group-delay length. Introducing the variable $\xi = t - z/V$ we obtain after substituting (4) in (3)

$$\frac{d\mathscr{E}_{0s}}{d\xi} = -\varkappa_s \frac{Vc_s}{V - c_s} \mathscr{E}_{0i} \sin \varphi_0, \quad \frac{d\mathscr{E}_{0i}}{d\xi} = \varkappa_i \frac{Vc_i}{V - c_i} \mathscr{E}_{0s} \sin \varphi_0,$$

(5)

whence

$$\mathscr{E}_{0s}^2 = a_s (C_1 - \cos \varphi_0), \quad \mathscr{E}_{0i}^2 = a_i (C_2 - \cos \varphi_0),$$

(6)

where C_1 and C_2 are the integration constants, and

$$a_s = -\frac{2\hbar}{\lambda} \frac{Vc_s}{V - c_s} \varkappa_s, \quad a_i = \frac{2\hbar}{\lambda} \frac{Vc_i}{V - c_i} \varkappa_i .$$

(7)

From (5) and (6) we have an equation for φ_0:

$$\frac{d\varphi_0}{d\xi} = \sqrt{b (C_1 - \cos \varphi_0)(C_2 - \cos \varphi_0)},$$

(8)

where $b = (\lambda^2 / \hbar^2) a_i a_s$. Introducing a new variable $x = \tan(\varphi_0 / 2)$, we obtain (8) in the form

$$\int_{x_0}^{x} \frac{dx}{\sqrt{b [C_1 - 1 + (C_1 + 1) x^2][C_2 - 1 + (C_2 + 1) x^2]}} = \frac{1}{2} \int_{\xi_0}^{\xi} d\xi.$$

(9)

The integral (9) contains all the soliton solutions of the system (3). Specifying the different C_1 and C_2, we obtain different classes of SRS solitons. From the requirement that (6) be positive and from the form of the expressions (7) for a_s, a_i ($W^{eq} < 0$) follow limitations on the dispersion characteristics of the medium, for which the existence of a given class of solitons is possible. We note that the quantities C_1 and C_2 are connected with the initial values of the fields and of the polarizations.

---

*We note that here and below we neglect the energy transfer to the anti-Stokes and higher components, since intense fields of the exciting and first-Stokes radiation are specified at the input to the medium, and the fields of the other components are triggered by noise.

114

Thus, an important role in the formation of solitons is played by group-delay effects. It should be noted in this connection that, as shown in [5], allowance for dispersion in the first-order approximation to which Eqs. (2) correspond is fully adequate for problems of optics of ultrashort pulses.

3. SOLITON SOLUTIONS OF SRS EQUATIONS

We consider now soliton regimes that correspond to different relations between the group velocities of the interacting waves (normal and anomalous dispersions) and to different initial populations of the working levels of the molecules of the medium.

I. We obtain the solitons corresponding to the cases $C_1 = C_2 = \pm 1$.

a) Let $C_1 = C_2 = 1$. It follows then from (6) and (7) that the velocity V should lie in the interval $c_s < V < c_i$ (i.e., a_i, $a_s > 0$). The condition $c_s < c_i$ which arises in this case corresponds to the case of anomalous dispersion [7]. We integrate (9) with the initial conditions $x_0 = 0$, $\xi_0 = -\infty$, which corresponds at the given choice of constants C_1 and C_2 to $\mathscr{E}_i(-\infty) = \mathscr{E}_s(-\infty) = 0$ and $\varphi_0(-\infty) = 0$, i.e., at the initial instant of the interaction of the fields the medium is at equilibrium. Then

$$\varphi_0(\xi) = 2\left[\frac{\pi}{2} + \mathrm{arctg}\,(\sqrt{b}\,\xi)\right],\ \theta = \varphi(\infty) - \varphi(-\infty) = 2\pi, \tag{10}$$

$$\mathscr{E}_{0s}^2 = \frac{2a_s}{1 + (\sqrt{b}\,\xi)^2},\quad \mathscr{E}_{0i}^2 = \frac{2a_i}{1 + (\sqrt{b}\,\xi)^2},\quad \tau^{-1} = \sqrt{b}, \tag{11}$$

which agrees with the solution obtained for (3) in [8] by another method.

b) Let $C_1 = C_2 = -1$. The soliton solutions are possible if the condition $c_i < V < c_s$ (a_i, $a_s < 0$) is satisfied, corresponding to the region of normal dispersion. Integrating (9) and $x_0 = 0$, $\xi_0 = 0$, we obtain $\varphi_0 = \sqrt{b}\xi$. From (6) we have

$$\mathscr{E}_{0s}^2 = -\frac{2a_s}{1 + (\sqrt{b}\,\xi)^2},\quad \mathscr{E}_{0i}^2 = -\frac{2a_i}{1 + (\sqrt{b}\,\xi)^2}, \tag{12}$$

with $\varphi_0(-\infty) = -\pi$, $\varphi_0(\infty) = \pi$, and $\theta = 2\pi$. It follows therefore that at $t = -\infty$ the considered pair of levels should have an inverted population $[W_{-\infty} = W^{eq}\cos\varphi_0(-\infty)]$. Thus, the propagation of 2π pulses of SRS with Lorentz shape in a medium with normal dispersion is possible only under conditions of inverted initial populations.

II. We consider now the cases $C_1 = C_2 = C > 1$, $C_1 = C_2 = C < -1$.

a) Let $C > 1$. Just as in the case of class I, we obtain the condition $c_s < V < c_i$ (a_i, $a_s > 0$), which is satisfied in the region of anomalous dispersion. According to (9) we have

$$\varphi_0(\xi) = 2\,\mathrm{arctg}\left[\sqrt{\frac{C-1}{C+1}}\,\mathrm{tg}\left(\frac{1}{2}\sqrt{b(C^2-1)}\,\xi\right)\right], \tag{13}$$

whence

$$\mathscr{E}_{0s}^2 = \frac{a_s(C^2-1)}{C + \cos(\sqrt{b(C^2-1)}\,\xi)},\quad \mathscr{E}_{0i}^2 = \frac{a_i(C^2-1)}{C + \cos(\sqrt{b(C^2-1)}\,\xi)}. \tag{14}$$

The solitons (14) constitute an unbounded periodic sequence of pulses with period $T = 2\pi/\sqrt{b(C^2-1)}$. The overlap integral over one period is equal to

$$\theta_T = \frac{\lambda}{\hbar}\int_0^T \mathscr{E}_{0i}\mathscr{E}_{0s}d\xi = 2\pi,$$

i.e., the solitons (14) are an infinite train of five pulses (we shall call them solitons of the trigonometric type). We note that in the present analysis we assume that there are no relaxation processes ($T_1 = T_2 = \infty$). Therefore in case (a) it is necessary to have $T \ll T_2$. In a real situation the influence of the finite T_2 should lead to a gradual noncoherent dephasing of the scatterers and consequently to termination of the train.

b) Let $C < -1$. From (6) and (7) follows here the condition $c_i < V < c_s$ (a_i, $a_s < 0$), which is satisfied in the case of normal dispersion. The solution (9) leads to expressions (13) and (14). The solitons make up an infinite train of 2π pulses of the trigonometric type.

III. We consider the case $C_1 > 1$, $C_2 > 1$ with $C_1 \neq C_2$. It follows from (6) that $c_s < V < c_i$ is the case of anomalous dispersion.

a) Let $C_1 > C_2$. We introduce the parameters $\alpha^2 = (C_1 - 1)(C_1 + 1)$, $\beta^2 = (C_2 - 1)(C_2 + 1)$ and make the change of variables $x = \beta \tan \psi$. Integration of (9) with the initial conditions $x_0 = 0$ ($\psi_0 = 0$), $\xi_0 = 0$ yields

$$F\left(\psi, \frac{\sqrt{\alpha^2 - \beta^2}}{\alpha}\right) = \frac{1}{2}\sqrt{b(C_1 - 1)(C_2 + 1)}\,\xi,$$

where $F(\psi, k)$ is an elliptic integral of the first kind. Hence

$$\varphi_0(\xi) = 2\,\mathrm{arctg}\left\{\beta\,\mathrm{tn}\left[\frac{1}{2}\sqrt{b(C_1 - 1)(C_2 + 1)}\,\xi\right]\right\}, \tag{15}$$

$$\mathscr{E}_{0s}^2 = a_s\,\frac{(C_1 - 1)(C_2 + 1) - 2(C_1 - C_2)\,\mathrm{sn}^2\left(\frac{1}{2}\sqrt{b(C_1 - 1)(C_2 + 1)}\,\xi\right)}{C_2 + 1 - 2\,\mathrm{sn}^2\left(\frac{1}{2}\sqrt{b(C_1 - 1)(C_2 + 1)}\,\xi\right)}, \tag{16}$$

$$\mathscr{E}_{0i}^2 = a_i\,\frac{C_2^2 - 1}{C_2 + 1 - 2\,\mathrm{sn}^2\left(\frac{1}{2}\sqrt{b(C_1 - 1)(C_2 + 1)}\,\xi\right)},$$

where sn and tn are elliptic functions. The solutions (16) constitute an infinite periodic train with period $T = 4K/\sqrt{b(C_1 - 1)(C_2 + 1)} = 2\tau K$, where $K(p)$, $p^2 = 2(C_1 - C_2)[(C_1 - 1)(C_2 + 1)]^{-1}$ is a complete elliptic integral of the first kind. It is easy to obtain $\theta_T = \varphi_0(\tau K) - \varphi_0(-\tau K) = 2\pi$, i.e., the solitons (16) are infinite trains of 2π pulses (we shall call them solitons of the elliptic type).

b) Let $C_1 < C_2$. In this case the solutions are obtained in analogy with case (a):

$$\varphi_0(\xi) = 2\,\mathrm{arctg}\left\{\alpha\,\mathrm{tn}\left[\frac{1}{2}\sqrt{b(C_1 + 1)(C_2 - 1)}\,\xi\right]\right\}, \tag{17}$$

$$\mathscr{E}_{0s}^2 = \frac{a_s(C_1 - 1)}{C_1 + 1 - 2\,\mathrm{sn}^2\left(\frac{1}{2}\sqrt{b(C_1 + 1)(C_2 - 1)}\,\xi\right)},$$

$$\mathscr{E}_{0i}^2 = a_i\,\frac{(C_1 + 1)(C_2 - 1) - 2(C_2 - C_1)\,\mathrm{sn}^2\left(\frac{1}{2}\sqrt{b(C_1 + 1)(C_2 - 1)}\,\xi\right)}{C_1 + 1 - 2\,\mathrm{sn}^2\left(\frac{1}{2}\sqrt{b(C_1 + 1)(C_2 - 1)}\,\xi\right)} \tag{18}$$

and are infinite trains of 2π pulses of elliptic type with period $T = 4K/\sqrt{b(C_1 + 1)(C_2 - 1)} = 2K\tau$.

IV. We consider the case $C_1 < -1$, $C_2 < -1$, $C_1 \neq C_2$. In this case $c_i < V < c_s$ is a region of normal dispersion. It is easily seen that the solution comprises infinite trains of 2π which take the form (15) and (16) in the case $C_1 > C_2$ and (17) and (18) in the case $C_1 < C_2$.

V. In analogy with classes III and IV we consider the following cases:

a) $C_1 > 1$, $C_2 < -1$, wherein the conditions $V > c_s$, $V > c_i$ ($a_s > 0$, $a_i < 0$) must be satisfied, and b) $C_1 < -1$, $C_2 > 1$, in which the condition $V < c_s$, $V < c_i$ ($a_s < 0$, $a_i > 0$) must be satisfied. In both cases no restrictions are imposed on the dispersion properties of the medium. The solitons are infinite periodic trains of 2π pulses, described in case (a) by expressions (17) and (18) and in case (b) by expressions (15) and (16).*

VI. An investigation of the solutions (9) shows that in the cases $C_1 \neq \pm 1$, $C_2 = 1$; $C_1 = 1$, $C_2 \neq \pm 1$ there are no soliton solutions of physical meaning at $\xi > 0$.

VIIa. Let $C_1 < -1$, $C_2 = -1$ ($c_i < V < c_s$ is the region of normal dispersion). In this case the integration of (9) at $x_0 = 0$, $\xi_0 = 0$ yields

$$\varphi_0(\xi) = 2\,\mathrm{arctg}\left[\sqrt{\frac{C_1 - 1}{C_2 + 1}}\,\mathrm{sh}\left(\frac{1}{2}\sqrt{-2b(C_1 + 1)}\,\xi\right)\right]. \tag{19}$$

For the fields we obtain the following solutions (we shall call them solitons of hyperbolic type)

$$\mathscr{E}_{0s}^2 = \frac{a_s(C_1 - 1)\,\mathrm{ch}^2(\xi/\tau_1)}{1 + \left(\frac{C_1 - 1}{C_1 + 1}\right)^2\,\mathrm{sh}^2\frac{\xi}{\tau}}, \qquad \mathscr{E}_{0i}^2 = -\frac{2a_i}{1 + \left(\frac{C_1 - 1}{C_1 + 1}\right)^2\,\mathrm{sh}^2(\xi/\tau)},$$

$$\tau^{-1} = \frac{1}{2}\sqrt{-2b(C_1 + 1)}. \tag{20}$$

Here $\varphi_0(-\infty) = -\pi$, $\varphi_0(\infty) = \pi$, and $\theta = 2\pi$. The fields (20) constitute 2π pulses, and at $t = -\infty$ the medium is inverted [$\varphi(-\infty) = -\pi$].

---

*In the particular case $c_i = c_s$ we also obtained solitons of this class in [9].

116

VIIb. In the case $C_1 = -1$, $C_2 < -1$ ($c_i < V < c_s$) we have

$$\varphi_0(\xi) = 2\operatorname{arctg}\left[\sqrt{\frac{C_2 - 1}{C_2 + 1}}\,\operatorname{sh}\frac{\xi}{\tau}\right], \qquad \tau^{-1} = \frac{1}{2}\sqrt{-2b(C_2 + 1)}, \tag{21}$$

$$\mathscr{E}_{0s}^2 = -\frac{2a_s}{1 + \left(\frac{C_2 - 1}{C_2 + 1}\right)^2 \operatorname{sh}^2\frac{\xi}{\tau}}, \qquad \mathscr{E}_{0i}^2 = \frac{a_i(C_2 - 1)\operatorname{ch}^2\frac{\xi}{\tau_2}}{1 + \left(\frac{C_2 - 1}{C_2 + 1}\right)^2 \operatorname{sh}^2\frac{\xi}{\tau}}. \tag{22}$$

Just as in case (a), $\varphi_0(-\infty) = -\pi$, $\varphi_0(\infty) = \pi$, $\theta = 2\pi$, i.e., the solutions (22) are 2π pulses.

VIIc. In the cases $C_1 > 1$, $C_2 = -1$ ($V > c_s$, $V > c_i$ – arbitrary dispersion) and $C_1 = -1$, $C_2 > 1$ ($V < c_s$, c_i – arbitrary dispersion) the soliton solutions are described, respectively, by the functions (19), (20) and (21), (22).

VIII. We consider the cases: a) $C_1 = 1$, $C_2 = -1$ ($V > c_s$, $V > c_i$ – arbitrary dispersion); b) $C_1 = -1$, $C_2 = 1$ ($V < c_s$, $V < c_i$ – arbitrary dispersion). In case a) we have

$$\varphi_0(\xi) = 2\operatorname{arctg}\left[\exp\left(\frac{\xi}{\tau}\right)\right], \qquad \tau = 1/\sqrt{b},$$

$$\mathscr{E}_{0s}^2 = \frac{2a_s \exp(2\xi/\tau)}{1 + \exp(2\xi/\tau)}, \qquad \mathscr{E}_{0i}^2 = -\frac{2a_i}{1 + \exp(2\xi/\tau)}. \tag{23}$$

In case (b) the solutions are obtained from (23) by making the substitutions $a_s \to -a_s$, $a_i \to -a_i$, $\mathscr{E}_{0s} \leftrightarrow \mathscr{E}_{0i}$. It follows from (23) that $\varphi_0(-\infty) = 0$, $\varphi_0(\infty) = \pi$, $\theta = \pi$. Thus, the fields (23) are π pulses.

IX. In cases when one of the constants is less than unity in absolute value, and the other is not equal to unity, the produced soliton regimes correspond to rotation angles of the polarization vector of the medium, bounded by individual parts of the interval $(-\pi, \pi)$. For example, at $C_1 = C_2 = C$ ($0 \le C < 1$), when $c_i < V < c_s$ – case of normal dispersion, we have

$$\mathscr{E}_{0s}^2 = -\frac{a_s(1 - C^2)}{C + \operatorname{ch}(\xi/\tau)}, \qquad \mathscr{E}_{0i}^2 = -\frac{a_i(1 - C^2)}{C + \operatorname{ch}(\xi/\tau)},$$

$$|\varphi_0| \le 2\operatorname{arctg}(1/a), \qquad \tau^{-1} = \sqrt{b(1 - C^2)}, \qquad a^2 = (1 + C)/(1 - C).$$

On the other hand if $c_s < V < c_i$, then

$$\mathscr{E}_{0s}^2 = \frac{a_s(1 - C^2)}{\operatorname{ch}(\xi/\tau + \alpha) - C}, \qquad \mathscr{E}_{0i}^2 = \frac{a_i(1 - C^2)}{\operatorname{ch}(\xi/\tau + \alpha) - C}, \qquad \alpha = \ln\frac{1 + a}{1 - a}.$$

In the first case $\theta = 4\tan^{-1}(1/a)$ and in the second case $\theta = 2\pi - \tan^{-1}(1/a)$. At $0 \le C_1 < 1$, $C_2 > 1$ we have for $V < c_i$, c_s

$$\varphi_0(\xi) = 2\operatorname{arctg}\left(\frac{a_1 a_2}{\sqrt{a_1^2 + a_2^2}}\,\frac{\operatorname{sh}\xi/\tau}{\operatorname{dn}(\xi/\tau)}\right), \qquad |\varphi_0| \le 2\operatorname{arctg} a_1,$$

and at $c_s < V < c_i$

$$\varphi_0(\xi) = 2\operatorname{arctg}\left[\frac{a_1}{\operatorname{cn}(\xi/\tau)}\right],$$

where $a_1^2 = \frac{1 - C_1}{1 + C_1}$, $a_2^2 = \frac{C_2 - 1}{C_2 + 1}$, $\tau^{-1} = \sqrt{2|b|(C_2 - C_1)}$. The solitons are infinite periodic trains of elliptic type with period $4K(a_1/\sqrt{a_1^2 + a_2^2})$ and with areas $\theta_T = 4\tan^{-1} a_1$ in the case of arbitrary dispersion and $\theta_T = 2\pi - \tan^{-1} a_1$ in the case of anomalous dispersion.

We note that the analysis presented here can be applied to the case of two-photon resonant absorption of pulses with unequal frequencies. For this process, the virtual level is located between the working levels of the molecule transition, i.e., $\omega_i + \omega_s = \omega_V$. It is easily seen that a description of two-photon absorption can be obtained by formally replacing ω_s by $(-\omega_s)$ in Eqs. (2) and (5). Thus, all the obtained classes of the soliton solutions exist also in the case of two-photon absorption, with the obvious change of the conditions imposed on the dispersion properties of the medium. We indicate also that the analysis that follows (Sections 4 and 5) is valid also for the case of two-photon absorption.

4. INFLUENCE OF RELAXATION ON SOLITON PROPAGATION

The results obtained above pertained to the case of total absence of relaxation ($T_1 = T_2 = \infty$). We shall show that allowance for finite T_1 and T_2 leads, just as in the case of the ordinary effect of self-induced transparency [2], to gradual small changes of the soliton pulses in the course of their propagation. The analysis presented below is valid both for single pulses and for each of the T_2/T periods of the train pulses. With the aid of (2) we easily find that the values of the energy of the pump pulses and of the Stokes wave passing through a unit cross-sectional area [$\varepsilon(\omega)$ is the dielectric constant of the medium]

$$\Omega_i = \frac{c\sqrt{\varepsilon(\omega_i)}}{8\pi} \int_{\delta_1}^{\delta_2} \mathscr{E}_i^2(z,t)\,dt, \qquad \Omega_s = \frac{c\sqrt{\varepsilon(\omega_s)}}{8\pi} \int_{\delta_1}^{\delta_2} \mathscr{E}_s^2(z,t)\,dt$$

satisfy the equations

$$\frac{1}{\omega_i \varepsilon(\omega_i)}\frac{d\Omega_i}{dz} = -\frac{1}{\omega_s \varepsilon(\omega_s)}\frac{d\Omega_s}{dz} = -c^2\frac{\hbar\mu_0 N_V}{8\pi}\int_{\delta_1}^{\delta_2}\frac{W(z,t)-W^{eq}}{T_1}\,dt,$$

where $\delta_1 = -\infty$, $\delta_2 = \infty$ in the case of single pulses and $\delta_1 = -\delta_2 = -T/2$ in the case of trains (T is the period of the train). These equations show that relaxation leads to additional transfer of energy from one wave to the other. We use next the relation $\int_{\delta_1}^{\delta_2}\{[u^2(z,t)+v^2(z,t)]/T_2 + W[W(z,t)-W^{eq}]/T_1\}\,dt = 0$, which can be easily obtained from (2) by considering the change of the length of the Bloch "vector" with components u, v, and W on account of relaxation. Then

$$\frac{1}{\omega_i \varepsilon(\omega_i)}\frac{d\Omega_i}{dz} = -\frac{1}{\omega_s \varepsilon(\omega_s)}\frac{d\Omega_s}{dz} = c^2\frac{\hbar\mu_0 N_V}{8\pi W^{eq}}\int_{\delta_1}^{\delta_2}\left\{\frac{u^2(z,t)+v^2(z,t)}{T_2} + \frac{[W(z,t)-W^{eq}]^2}{T_1}\right\}\,dt.$$

This formula is valid for pulses of arbitrary type (not necessarily soliton pulses) and for arbitrary ratios of their durations and relaxation times. We turn to solutions of the soliton type and use the smallness of the ratios τ/T_1, τ/T_2, where the values of τ, determined above for each type of soliton, have the meaning of the characteristic scale of their duration (in the case of trains the same role is played by the ratios T/T_1, T/T_2). We then have

$$\frac{1}{\omega_i \varepsilon(\omega_i)}\frac{d\Omega_i}{dz} = -\frac{1}{\omega_s \varepsilon(\omega_s)}\frac{d\Omega_s}{dz} = c^2\frac{\hbar\mu_0 N_V W^{eq}}{4}\left(\frac{\tau}{T_1}+\frac{\tau}{T_2}\right).$$

The obtained expression shows that the rate of change of the soliton energy is a small quantity $\sim\tau(T_1^{-1}+T_2^{-1})$, i.e., allowance for small relaxation terms leads to a weak influence of these terms on the passage of the solitons (in this sense, our problem does not differ from other problems in the theory of ultrarelativistic pulses [10]). The relaxation effects lead at $W^{eq} < 0$ to a gradual damping of the pump and to an increase of the intensity of the Stokes wave. This is due both to the increase of the population of the lower level with a characteristic time T_1, and to the attenuation of the polarization of the medium with a characteristic time T_2. Conversion of one wave into another proceeds, however, much more slowly than in the case of an exponential regime of nonstationary SRS [5]. It is easy to show that the critical length over which the relaxation effects disrupt the soliton regime is given by the expression

$$z_c \sim \frac{2a_i\tau}{c\hbar\mu_0 N_V \omega_i |W^{eq}|\sqrt{\varepsilon(\omega_i)}}\left(\frac{\tau}{T_1}+\frac{\tau}{T_2}\right)^{-1}.$$

It should be noted that in the absence of relaxation ($T_1 = T_2 = \infty$) $z_c \to \infty$. By virtue of the specifics of the combination interaction, the relaxation leads not to a gradual absorption of the solitons, as in the case of resonant passage [2], but to a gradual amplification of one wave at the expense of the other. Just as in the case of the ordinary effect of self-induced transparency [2], the influence of the relaxation processes can be compensated with the aid of weak focusing of the beams in the medium.

Thus, the SRS solitons, as well as all other waves of this type [11], are intermediate self-similar asymptotics that exist until the relaxation effects come into play. To prove their physical realizability it is necessary to vertify that the soliton waves are stable with respect to small perturbations, sources of which are always present in experiment (inhomogeneity of the composition of the medium, fluctuations of the thermodynamic quantities, etc.).

5. STABILITY OF SRS SOLITONS

We shall regard SRS solitons as stable if small perturbations of the fields do not begin to increase with time and upset their stationary character. Introducing the quantities

$$\varphi_{i,s}(z,t) = \frac{\lambda}{\hbar} \int \mathscr{E}_{i,s}^2(z,t)\,dt, \qquad \varphi(z,t) = \frac{\lambda}{\hbar} \int \mathscr{E}_s(z,t)\,\mathscr{E}_i(z,t)\,dt, \tag{24}$$

we obtain with the aid of (3)

$$\frac{\partial \varphi_s}{\partial z} + \frac{1}{c_s} \frac{\partial \varphi_s}{\partial t} = 2\varkappa_s(\cos\varphi - C_1),$$

$$\frac{\partial \varphi_i}{\partial z} + \frac{1}{c_i} \frac{\partial \varphi_i}{\partial t} = -2\varkappa_i(\cos\varphi - C_2), \qquad \left(\frac{\partial \varphi}{\partial t}\right)^2 = \frac{\partial \varphi_s}{\partial t} \frac{\partial \varphi_i}{\partial t}. \tag{25}$$

The stationary pulses obtained in Section 3 correspond to stationary solutions of the system (25):

$$\varphi_{0s} = \frac{\lambda}{\hbar} \int_{\xi_0}^{\xi} \mathscr{E}_{0s}^2(\xi)\,d\xi, \qquad \varphi_{0i} = \frac{\lambda}{\hbar} \int_{\xi_0}^{\xi} \mathscr{E}_{0i}^2(\xi)\,d\xi, \qquad \varphi_0(\xi) = \frac{\lambda}{\hbar} \int_{\xi_0}^{\xi} \mathscr{E}_{0s}(\xi)\,\mathscr{E}_{0i}(\xi)\,d\xi.$$

An investigation of the stability of the solitons $\mathscr{E}_{0s}$, $\mathscr{E}_{0i}$ is equivalent to an investigation of the stability of the quantities φ_{0s}, φ_{0i}, φ_0. We consider the behavior of small perturbations near $\varphi_0(\xi)$, $\varphi_{0s}(\xi)$, $\varphi_{0i}(\xi)$

$$\varphi_{s,i}(z,t) = \varphi_{0s,i}(\xi) + f_{s,i}(z,t), \qquad \varphi(z,t) = \varphi_0(\xi) + f(z,t), \tag{26}$$

where $|f_s| \ll |\varphi_{0s}|$, $|f_i| \ll |\varphi_{0i}|$, $|f| \ll |\varphi_0|$. According to (25), the system of equations describing the evolution of the small perturbations is of the form

$$\frac{\partial f_{s,i}}{\partial z} + \frac{1}{c_{s,i}} \frac{\partial f_{t,i}}{\partial t} = \mp 2\varkappa_{s,i} f \sin\varphi_0, \qquad \frac{\partial f}{\partial t} = \frac{1}{2}\left(\frac{\mathscr{E}_{0i}}{\mathscr{E}_{0s}} \frac{\partial f_s}{\partial t} + \frac{\mathscr{E}_{0s}}{\mathscr{E}_{0i}} \frac{\partial f_i}{\partial t}\right). \tag{27}$$

In the general case an investigation of this system entails great difficulties. It simplifies, however, for solitons of classes ($C_1 = C_2 = C$), for which it is possible with the aid of (24) to connect directly the perturbations f, f_s, and f_i: $f = \frac{1}{2}\left(\frac{a_i}{a_s} f_s + \frac{a_s}{a_i} f_i\right)$. Using this relation in place of the last equation in (27) and changing over to the variables $\xi = t - z/V$ and $p = t$, we obtain

$$\frac{\partial^2 f}{\partial \xi^2} + (\gamma_1 + \gamma_2)\frac{\partial^2 f}{\partial \xi\,\partial p} + \gamma_1\gamma_2 \frac{\partial^2 f}{\partial p^2} - \sqrt{b}\left[\frac{\partial f}{\partial \xi} + \frac{1}{2}(\gamma_1 + \gamma_2)\frac{\partial f}{\partial p}\right]\sin\varphi_0 - \sqrt{b}\,f\frac{\partial \varphi_0}{\partial \xi}\cos\varphi_0 = 0, \tag{28}$$

where

$$\gamma_1 = V/(V - c_s), \qquad \gamma_2 = V/(V - c_i).$$

1. We seek a solution of (28) in the form

$$f(\xi, p) = \sum_{l=0}^{\infty} f_l(\xi)\exp(\alpha_l p), \tag{29}$$

where $f_l(\xi)$ are assumed to be bounded functions at $|\xi| < \infty$ and $f(\xi, 0) = \sum_{l=0}^{\infty} f_l(\xi)$ gives the distribution of the perturbation at the initial instant of time $p = 0$. If the real part of at least one of the exponents α_l turns out to be positive, then by virtue of (29) the soliton regime is unstable. If the conditions Re $\alpha_l < 0$ is satisfied for all l, then the soliton regime is asymptotically stable. The perturbation tends then to attenuate with increasing p, and the system tends to return to the soliton regime. If Re $\alpha_l \le 0$, then the soliton regime is stable, but it has no asymptotic stability, since the system does not return to the initial state but is located close to it under the influence of a small perturbation even as $p \to \infty$. Substituting (29) in (28) and introducing a new function $K_l(\xi)$ according to

$$f_l(\xi) = K_l(\xi)\exp\left\{-\frac{1}{2}\int[\alpha_l(\gamma_1 + \gamma_2) - \sqrt{b}\sin\varphi_0]\,d\xi\right\},$$

we obtain a "Schrödinger equation"

$$-\frac{d^2 K_l}{d\xi^2} + \left[\frac{1}{4}b\sin^2\varphi_0 + \frac{1}{2}\sqrt{b}\frac{d\varphi_0}{d\xi}\cos\varphi_0 - \lambda_l\right]K_l = 0,$$

119

$$\lambda_l = -\frac{1}{4}(\gamma_1 - \gamma_2)^2 \alpha_l^2 \tag{30}$$

with a "potential"

$$U(\xi) = \frac{1}{4} b \sin^2 \varphi_0 + \frac{1}{2}\sqrt{b}\,\frac{d\varphi_0}{d\xi}\cos\varphi_0$$

and with "energy" eigenvalues λ_l. The spectrum λ_l of Eq. (30) is bounded from below, starting with a certain λ_0 corresponding to one state. According to the properties of the potential $U(\xi)$ the initial part of the spectrum can be discrete. We number the λ_i in the discrete spectrum in increasing order $\lambda_0 < \lambda_1 < \lambda_2 < \dots$. Then the index of a given λ_i will simultaneously give the number of zeros corresponding to the eigenfunction K_i of Eq. (30). The eigenfunction of the ground state K_0 should have no zeros. We shall show that in the case considered here the ground state corresponds to $\lambda_0 = 0$. The bounded solution of Eq. (30) with $\lambda_0 = 0$ is of the form

$$K_0(\xi) = \exp\left\{\frac{1}{2}\sqrt{b}\int \sin\varphi_0\, d\xi\right\}.$$

Using (8) we easily obtain

$$K_0(\xi) = \left(D\,\frac{d\varphi_0}{d\xi}\right)^{1/2}, \tag{31}$$

where D is a constant. Substituting the concrete expressions of $\varphi_0(\xi)$ for class I [see (10)], class II [see (13)], and class IX, we easily verify that the derivative $d\varphi_0/d\xi$ does not vanish anywhere. Consequently $\lambda_0 = 0$ corresponds to the ground state, and all the remaining eigenvalues are positive, i.e., $\lambda_l > 0$, $l = 1, 2, \dots$. Taking now into account the connection between λ_l and α_l, we find ultimately that the exponents α_l at $l \geq 1$ are purely imaginary, and $\alpha_0 = 0$. Thus, Eq. (28) for the perturbation of the overlap integral does not have exponentially growing solutions, i.e., the solitons of classes I, II, and IX are stable, although they have no asymptotic stability.

2. The absence of asymptotic stability may mean the existence of weakly growing perturbations. It is therefore necessary to investigate the influence of perturbations of this type on the soliton regimes. We shall show that there exists a perturbation that increases linearly with time p, in the form

$$f(\xi, p) = A(\xi) + B(\xi)\,p. \tag{32}$$

Substituting (32) in (28) we get

$$\frac{d^2 A}{d\xi^2} - \sqrt{b}\sin\varphi_0\frac{dA}{d\xi} - \sqrt{b}\,\frac{d\varphi_0}{d\xi}\cos\varphi_0 A + (\gamma_1 + \gamma_2)\frac{dB}{d\xi} - \frac{1}{2}\sqrt{b}\,(\gamma_1 + \gamma_2)B\sin\varphi_0 = 0, \tag{33}$$

$$\frac{d^2 B}{d\xi^2} - \sqrt{b}\sin\varphi_0\frac{dB}{d\xi} - B\sqrt{b}\,\frac{d\varphi_0}{d\xi}\cos\varphi_0 = 0. \tag{33a}$$

Differentiating now Eq. (8) twice with respect to ξ and comparing the obtained equation with (33a), we find that the bounded solution (33a) takes the form $B(\xi) = d\varphi_0/d\xi$. By direct substitution in (33) we can show that

$$A(\xi) = -B(\xi)(\gamma_1 + \gamma_2)\,\xi/2.$$

Thus, the perturbation that increases linearly with time is of the form

$$j(\xi, p) = \left[p - \frac{1}{2}(\gamma_1 + \gamma_2)\xi\right]\frac{d\varphi_0}{d\xi}. \tag{34}$$

Formally, the result obtained can be interpreted as the presence of instability of the solitons with respect to the perturbation (34). We shall show, however, using the approach used in [12] in the investigation of the stability of solitons in a medium with single-photon absorption, that such perturbations do not break up the SRS solitons. We write down the complete expression (26) for the perturbed overlap integral $\varphi(z, t)$ near the soliton solution $\varphi_0(\xi)$:

$$\varphi(z, t) = \varphi_0(\xi) + \nu\,[p - \delta\xi]\frac{d\varphi_0}{d\xi}, \tag{35}$$

where we introduced explicitly the small perturbation amplitude ν, and $\delta = \frac{1}{2}(\gamma_1 + \gamma_2)$. Expression (35) is a series expansion in the linear approximation in ν. Therefore, with the same accuracy, we have

$$\varphi(z, t) = \varphi_0\,[(\xi + \nu\,(p - \delta\xi))/\tau], \tag{36}$$

120

where we have introduced the duration τ of the unperturbed pulses (or the period in the case of a train):

$$\tau^{-1} = \begin{cases} \sqrt{b} & , \quad C = \pm 1, \\ \sqrt{b(C^2-1)}, & C > 1, C < -1, \\ \sqrt{b(1-C^2)}, & -1 < C < 1. \end{cases}$$

The possibility of representing (35) in the form (36) shows that the perturbed fields again constitute solitons, but now with a different velocity and duration compared with the initial ones. This can be easily verified by transforming the argument in (36) into $\varphi_0(z, t) = \varphi_0[(p - z/V')/\tau']$, where τ' is the duration and V' the velocity of the perturbed pulses

$$\tau'^{-1} = \tau^{-1}[1 + \nu(1-\delta)], \quad V' = V\left[1 + \frac{\nu}{1-\delta\nu}\right].$$

It can be shown that in first order of smallness in ν the quantities τ' and V' are connected by the same relation as the quantities τ and V:

$$\frac{1}{\tau'} = \begin{cases} \sqrt{b'} & , \quad C = \pm 1, \\ \sqrt{b'(C^2-1)}, & C > 1, C < -1, \\ \sqrt{b'(1-C^2)}, & -1 < C < 1, \end{cases}$$

where $b' = 4\varkappa_i\varkappa_s c_i c_s V'^2/(V'-c_s)(V'-c_i)$. This result means that the perturbed solitons pertain to the same type of 2π pulses as the unperturbed ones. Consequently, the perturbations (32), while increasing with time, do not destroy the solitons and merely change their durations and velocities, preserving at the same time the shape of the envelope of the pulse. Under the influence of the perturbations (32), one 2π pulse goes over continuously into another, and, just as in the case considered in Item 1 of Section 5, no return to the initial regime takes place.

It should be noted that an investigation of the stability of solitons is most important, since, by using Eq. (2), conclusions that they break up into solitons in the course of propagation through the medium can be drawn only in the particular case of "proportional" input pulses. Actually, from (2) we have, for example for the case $C_1 = C_2 = 1$,

$$\frac{d\varphi_s(z)}{dz} = -\frac{\lambda\omega_s c_s\mu_0 N_V W^{eq}}{2} 2\sin^2\frac{\varphi(z)}{2},$$

$$\frac{d\varphi_i(z)}{dz} = \frac{\lambda\omega_i c_i\mu_0 N_V W^{eq}}{2} 2\sin^2\frac{\varphi(z)}{2},$$

$$\frac{d\varphi(z)}{dz} = \frac{\lambda^2\mu_0 N_V W^{eq}}{2\hbar}\int_{-\infty}^{\infty}[c_i\omega_i\mathscr{E}_s^2 - c_s\omega_s\mathscr{E}_i^2]\sin\varphi(z,t')\,dt' - \frac{\lambda}{\hbar}\int_{-\infty}^{\infty}\left[\frac{1}{c_s}\mathscr{E}_i\frac{\partial\mathscr{E}_s}{\partial t} + \frac{1}{c_i}\mathscr{E}_s\frac{\partial\mathscr{E}_i}{\partial t}\right]dt,$$

where $\varphi_{i,s}(z)$, $\varphi(z)$ are determined in accordance with (24) with integration with respect to t from $-\infty$ to $+\infty$. The pulses whose amplitude are proportional, $\mathscr{E}_i(z,t)/\mathscr{E}_s(z,t) = $ const and $\varphi(0) = 2\pi m$, $m > 1$, satisfy these equations, and furthermore at arbitrary z: $\varphi(z) = 2\pi m$, $d\varphi_s(z)/dz = d\varphi_i(z)/dz = 0$, i.e., $\varphi_{i,s}(z)$, $\varphi(z)$ do not change in the course of pulse propagation. By virtue of the uniqueness of the solution this means that pulses having on entering the medium the properties $\mathscr{E}_i(0,t)/\mathscr{E}_s(0,t) = $ const, $\varphi_0(0) = 2\pi m$, should remain proportional. However, according to Section 3 at $C_1 = C_2 = 1$ only pulses with $\varphi(0) = 2\pi$ are stationary. Consequently, proportional input pulses should break up in the course of propagation into m pairs of solitons (11) and remain proportional in the course of the breakup. For arbitrary input pulses, the question of the onset of the soliton regimes can be solved by a numerical analysis of (2).

In conclusion let us estimate the threshold amplitudes. For typical substances $\lambda \simeq 10^{-23}$-10^{-25} cm^3. According to the formulas presented above, assuming the soliton duration (the period of the train) to be smaller by one order of magnitude than T_2, we obtain

$$\mathscr{E}_i\mathscr{E}_s \gg \frac{2\pi\hbar}{\lambda(T_2/10)} \simeq \frac{10^2\text{-}10^4}{T_2}\left[\frac{V}{cm}\right]^2.$$

In typical situations $T_2 \sim 10^{-11}$ sec for liquids and $T_2 \sim 10^{-8}$-10^{-9} sec for gases. This yields for liquids $\sqrt{\mathscr{E}_i\mathscr{E}_s} \gtrsim 10^7$-$10^8$ V/cm and for gases $\sqrt{\mathscr{E}_i\mathscr{E}_s} \gtrsim 10^5$-$10^6$ V/cm.

121

6. CONCLUSIONS

The results show that besides the well-investigated quasistatic and nonstationary SRS regimes, in strong fields of ultrashort pulses there are produced unique soliton scattering regimes that are not accompanied by amplification. Depending on the initial conditions and on the dispersion characteristics of the scattering medium, soliton regimes of different classes can be realized. In particular, in the region of normal dispersion, for an initially equilibrium medium, the SRS solitons can constitute only periodic trains of 2π pulses. Solitons in the form of solitary 2π pulses of finite duration can be realized under the same initial conditions only in the region of anomalous dispersion. Similar soliton regimes in the region of normal dispersion are realized only upon inversion of the initial population of the levels of the working transition, and constitute solitons of Lorentz shape or 2π pulses of hyperbolic type. Under arbitrary dispersion conditions, regimes are also possible in the form of periodic trains of 2π pulses of elliptic and trigonometric type. Solitons with Lorentz shape, trains of trigonometric type, and a special type of solitary solitons with $\theta < 2\pi$ (class IX) are stable, thus pointing to the feasibility of their practical realization (see the estimate in Section 5). Relaxation processes lead to a gradual slow departure from the soliton regime of scattering.

We note also that soliton regimes of SRS can be of great importance in the analysis of the detailed temporal structure of the emission of Raman lasers, since the fields produced in such systems have intensive fluctuation bursts of small duration. Observation of soliton regimes "in pure form" is best carried out in low-pressure gases ~0.1 atm at input laser-pulse durations ~1-10 nsec.

The authors thank S. A. Akhmanov for a discussion of the results and for helpful remarks, and G. M. Makhviladze for a discussion of the question of the stability of soliton solutions.

LITERATURE CITED

1. I. A. Poluéktov, Yu. M. Popov, and V. S. Roitberg, Usp. Fiz. Nauk, 114:97 (1974).
2. S. L. McCall and E. L. Hahn, Phys. Rev. Lett., 18:908 (1967); Phys. Rev., 183:457 (1969).
3. É. M. Belenov and I. A. Poluéktov, Zh. Éksp. Teor. Fiz., 56:1407 (1969).
4. N. Tan-no, K. Yokoto, and H. Inaba, Phys. Rev. Lett., 29:1211 (1972).
5. S. A. Akhmanov, K. N. Drabovich, A. P. Sukhorukov, and A. S. Chirkin, Zh. Éksp. Teor. Fiz., 59: 485 (1970).
6. T. M. Makhviladze and M. E. Sarychev, Preprint FIAN, No. 165 (1975); Zh. Éksp. Teor. Fiz., 71:896 (1976).
7. S. A. Akhmanov, K. N. Drabovich, A. P. Sukhorukov, and A. K. Shchednova, Zh. Éksp. Teor. Fiz., 62:525 (1972).
8. T. M. Makhviladze, M. E. Sarychev, and L. A. Shelepin, Preprint FIAN, No. 18 (1974); Zh. Éksp. Teor. Fiz., 69:499 (1975).
9. N. Tan-no, T. Shirahata, and K. Yokoto, Phys. Rev., A12:159 (1975).
10. P. G. Kryukov and V. S. Letokhov, Usp. Fiz. Nauk, 99:169 (1969).
11. G. I. Barenblatt, Preprint IPM, No. 52 (1975).
12. S. R. Barone, Lett. Nuovo Cimento, 3:156 (1972).

ANGULAR DISTRIBUTIONS OF STIMULATED RAMAN SCATTERING OF LIGHT

T. M. Makhviladze and M. E. Sarychev

We consider stimulated Raman scattering, given the pump fields and the first Stokes component of different spatial configuration (parallel beams, crossed beams, axial and spherically symmetrical field of the first Stokes component). It is shown that, depending on the experimental conditions, there can be realized either emission of the Townsend type, or two other types of emission whose intensity is unstable to changes of the geometric parameters and can have a complicated diffraction substructure. A comparative analysis is made of the various types of radiation. The theoretical results are compared with the available experimental data.

1. INTRODUCTION

The main peculiarity of stimulated Raman scattering of light (SRS) is the complicated angular distribution of the scattered radiation. The existence of cones of preferred propagation of the SRS radiation follows even from consideration of the power transferred to the scattered component by the molecular oscillator of the medium, which is excited near resonance by the interaction of the incident and scattered components of the field. The cone aperture angle follows then from the relations (the Townes rule [1])

$$\mathbf{k}_0 + \mathbf{k}_{-m} = \mathbf{k}_{-1} + \mathbf{k}_{-m+1}, \qquad \mathbf{k}_0 + \mathbf{k}_m = \mathbf{k}_{-1} + \mathbf{k}_{m+1}, \tag{1}$$

where $\mathbf{k}_{-m}$ is the wave vector of the field of the m-th Stokes [$(\mathbf{k}_m - m)$-th anti-Stokes] component, and $\mathbf{k}_0$ is the wave vector of the pump field. Conditions (1) can be obtained in more general form:

$$\mathbf{k}_n + \mathbf{k}_m = \mathbf{k}_{n-1} + \mathbf{k}_{m+1}, \tag{1a}$$

where n and m can assume positive and negative values. Expression (1a) is simply the synchronism condition in four-wave interaction. The conditions (1) are separated from (1a) because of the experimental situation wherein the SRS components are formed by a successive mechanism of energy transfer from the pump field. This simplified analysis presupposes that the fields of the scattered components are plane waves, and essentially does not take into account the real geometry of the active volume.

A detailed theory of the successive development of the SRS components, based on a solution of the three-dimensional equations of electrodynamics [2, 3], shows that the SRS fields are spherical waves, and the directions of the maximum intensity are determined from the relations [2]

$$\mathbf{k}_{-m} = m\mathbf{k}_{-1} - (m+1)\mathbf{k}_0, \qquad \mathbf{k}'_m = (m+1)\mathbf{k}_0 - m\mathbf{k}_{-1}. \tag{2}$$

An important aspect in this theory is the need for introducing into Maxwell's equations a phenomenological term that describes the bare radiation of the first Stokes component. It should be noted that the rules (2) are free of a shortcoming of the Townes rule — the possibility of noncoplanar combinations of vectors [2] — and yield emission angles that coincide with the Townes angles for coplanar combinations in (1).

Further experiments [4, 5] have revealed that in a liquid the conditions (1) and (2) describe only a small fraction of the produced angular distributions, which were named emissions of class I. The highest intensity, however, is possessed by the off-axis radiation, which does not obey these rules — emission of class II (rings of class II). In [4, 6] attempts were made to connect this radiation with the self-focusing phenomenon. It has also been proposed that this radiation arises also in the absence of self-focusing and is due to the always present paraxial Stokes radiation [7]. Arbatskaya and Sushchinskii [7] investigated experimentally the angular distributions of the higher Stokes and anti-Stokes components, applying to the input of the cell with the investigated substance, besides the pump, also the first Stokes component radiation excited beforehand in a second cell and directed strictly along the sample axis. Intensive radiation of class II was observed at angles that turned out to be in good agreement with the experimental results [4, 5], where no special measures were made to amplify the paraxial Stokes radiation.

To assess the role of the paraxial Stokes radiation, Section 2 is devoted to a theoretical analysis of the angular distributions of the second Stokes and first anti-Stokes SRS components, when the fields of the pump and of the first Stokes component are specified intense plane waves of constant amplitude and propagate along the sample axis. In Section 3 we consider the radiation of the higher Stokes components. In Sections 2 and 3, a comparison is made of the obtained angular distributions with the experiments [4, 5, 7]. In Sections 4-6 we investigate SRS in the fields of the pump and of the first Stokes component of varying spatial configuration. Such an analysis permits the various types of radiation produced in SRS to be treated in the same general manner, and allows a comparison of the singularities of their structure. The results obtained in Sections 4-6 can be verified in experiments in which a two-cell scheme is used. The present article is based on the author's earlier papers [8-11].

2. SRS IN PARALLEL PUMP AND FIRST STOKES COMPONENT BEAMS

We consider Raman scattering on one pair of levels of isotropic scattering molecules. We assume the fields produced at all the combination frequencies to be classical, and describe them with the aid of three-dimensional Maxwell's equations. The Hamiltonian of the scattering by an individual molecule takes the form

$$H = \hbar\omega_v\sigma^z - \frac{\lambda}{2}\Big[\sigma^+ \sum_{l=-\infty}^{\infty} (\mathbf{E}_l^- \mathbf{E}_{l-1}^+) + \sigma^- \sum_{l=-\infty}^{\infty} (\mathbf{E}_l^+ \mathbf{E}_{l+1}^-)\Big], \tag{3}$$

where σ^z, $\sigma^\pm$ are Pauli matrices and describe the transitions between the levels of the molecules; ω_V is the transition frequency; λ is the scattering matrix element; the field of the l-th component is defined by the expression

$$\mathbf{E}_l(\mathbf{r}, t) = \frac{1}{2}[\mathbf{E}_l^+(\mathbf{r}, t) + \mathbf{E}_l^-(\mathbf{r}, t)] = \frac{1}{2}[\varepsilon_l(\mathbf{r}, t)\exp(i\omega_l t) + \varepsilon_l^*(\mathbf{r}, t)\exp(-i\omega_l t)],$$

where $\varepsilon_l(\mathbf{r}, t)$ are the amplitudes; $\omega_l = \omega_0 + l\omega$; ω_0 is the pump-field frequency; $\omega = \omega_l - \omega_{l-1}$, with $l < 0$ corresponding to the l-th Stokes component and $l > 0$ to the l-th anti-Stokes component. The polarization at the frequency of the m-th component is equal to

$$\mathbf{P}_m^+ = \frac{\lambda}{2} N_V [\mathbf{E}_{m+1}^+ \langle\sigma^+\rangle + \mathbf{E}_{m+1}^- \langle\sigma^-\rangle], \tag{4}$$

where $\langle\ \rangle$ denotes averaging with the aid of the density matrix, and N_V is the molecule density. The averages $\langle\sigma^\pm\rangle$ are obtained from the equations of motion for the operators $\sigma^\pm$, σ^z:

$$\Big(\frac{d}{dt} + \frac{1}{T_2} - i\omega_v\Big)\langle\sigma^+\rangle = i\frac{\lambda}{\hbar}\langle\sigma^z\rangle \sum_{l=-\infty}^{\infty} (\mathbf{E}_l^+ \mathbf{E}_{l-1}^-),$$

$$\Big(\frac{d}{dt} + \frac{1}{T_2} + i\omega_v\Big)\langle\sigma^-\rangle = -i\frac{\lambda}{\hbar}\langle\sigma^z\rangle \sum_{l=-\infty}^{\infty} (\mathbf{E}_l^- \mathbf{E}_{l-1}^+), \tag{5}$$

$$\frac{d\langle\sigma^z\rangle}{dt} + \frac{\langle\sigma^z\rangle - \langle\sigma^z\rangle^{eq}}{T_1} = -\frac{\lambda i}{2\hbar} \sum_{l=-\infty}^{\infty} [(\mathbf{E}_l^- \mathbf{E}_{l-1}^+)\langle\sigma^-\rangle - (\mathbf{E}_l^+ \mathbf{E}_{l-1}^-)\langle\sigma^+\rangle],$$

where T_1 is the time of the longitudinal relaxation, T_2 is the time of the transverse relaxation, and $\langle\sigma^z\rangle^{eq}$ is the equilibrium value of $\langle\sigma^z\rangle$.

We consider next a quasistatic scattering regime in an approximation where the population difference $\langle\sigma^z\rangle = \langle\sigma^z\rangle^{eq} = W^{eq}$ is given. Then, solving the system (5), we obtain according to (4)

$$\mathbf{P}_m^+ = \Big[\frac{\chi}{\delta - i}\varepsilon_{m-1} \sum_{l=-\infty}^{\infty} (\varepsilon_l \varepsilon_{l-1}^*) + \frac{\chi}{\delta + i}\varepsilon_{m+1} \sum_{l=-\infty}^{\infty} (\varepsilon_l^* \varepsilon_{l-1})\Big]\exp(i\omega_m t), \tag{6}$$

where $\chi = \lambda^2 N_V W^{eq} T_2 / 2\hbar$, $\delta = T_2(\omega - \omega_V)$, and the amplitudes ε_l depend in the considered approximation only on $\mathbf{r}$ (a derivation of the expression (6) in a purely classical approach is given in [2]). The field $\mathbf{E}_m^+$ is now obtained from Maxwell's equations with polarization (6):

$$\Big(\text{grad div} - \Delta - \frac{1}{c_m^2}\frac{\partial^2}{\partial t^2}\Big)\mathbf{E}_m^+ = -\mu_0 \frac{\partial^2 \mathbf{P}_m^+}{\partial t^2}. \tag{7}$$

where $\mu_0 = 4\pi/c^2$ and c is the speed of light at the frequency ω_m. In the solution of Eqs. (7) we use the following assumption:

$$|\mathbf{E}_{l+1}| \ll |\mathbf{E}_l|, \quad l \geqslant 0,$$
$$|\mathbf{E}_{l-1}| \ll |\mathbf{E}_l|, \quad l \leqslant 0, \tag{8}$$

which is valid for a sequential mechanism of excitation of the components over not too long times, when the depletion of the pump can be neglected.

We consider now the radiation at the second Stokes and first anti-Stokes frequencies, assuming the pump field $\mathbf{E}_0$ and the field $\mathbf{E}_{-1}$ of the first Stokes component to be specified plane waves of constant intensity propagating along the axis of the active volume:

$$\boldsymbol{\varepsilon}_0 = \mathbf{a}_0 \exp\left(-i\mathbf{k}_0\mathbf{r}\right), \quad \boldsymbol{\varepsilon}_{-1} = \mathbf{a}_{-1} \exp\left(-i\mathbf{k}_{-1}\mathbf{r}\right),$$

where $\mathbf{k}_0 \parallel \mathbf{k}_{-1}$; $\mathbf{a}_0$, $\mathbf{a}_{-1}$ are constant amplitudes. In our case, which corresponds to a two-cell experiment [7], the intensity of the first Stokes component greatly exceeds the intensity of any other component. We shall assume that the active volume V is a cylinder (of length l and radius r_0) with axis parallel to $\mathbf{k}_0$ (the z axis). We consider next radiation in a narrow cone near the z axis. Inasmuch as in this cone the fields in the plane (x, y) vary much less than along the z axis, we can neglect in the equations (7) the term $\nabla \operatorname{div} \mathbf{E}_m^+$ and assume the field amplitudes to be scalar. Then, substituting (6) in (7) and taking the condition (8) into account, we obtain ($\mathscr{E}_{-1}$, $\mathscr{E}_{-2}$, a_0, a_{-1} now denote scalar amplitudes)

$$(\Delta + p^2)\mathscr{E}_1 = -\chi_1 \left[a_0^2 a_{-1}^* \exp\left\{-i\left(2\mathbf{k}_0 - \mathbf{k}_{-1}\right)\mathbf{r}\right\} + a_0 a_{-1}\mathscr{E}_{-2}^* \exp\left\{-i\left(\mathbf{k}_0 + \mathbf{k}_{-1}\right)\mathbf{r}\right\}\right] g(r),$$

$$(\Delta + q^2)\mathscr{E}_{-2} = -\chi_{-2}\left[a_0^* a_{-1}^2 \exp\left\{-i\left(2\mathbf{k}_{-1} - \mathbf{k}_0\right)\mathbf{r}\right\} + a_0 a_{-1}\mathscr{E}_1^* \exp\left\{-i\left(\mathbf{k}_0 + \mathbf{k}_{-1}\right)\mathbf{r}\right\}\right] g(r), \tag{9}$$

where

$$p^2 = k_1^2 + \chi_1 |a_0|^2 g(\mathbf{r}), \qquad q^2 = k_{-2}^2 + \chi_{-2}|a_{-1}|^2 g(\mathbf{r}),$$

$$k_1^2 = \omega_1^2/c_1^2, \qquad k_{-2}^2 = \omega_{-2}^2/c_{-2}^2, \qquad \chi_1 = \frac{\chi\mu_0\omega_1^2}{\delta - i}, \qquad \chi_{-2} = \frac{\chi\mu_0\omega_{-2}^2}{\delta + i}, \tag{10}$$

and the function $g(\mathbf{r})$ is equal to unity inside the scattering volume and to zero outside this volume. We note that in our problem the role of the bare sources is played by the input fields of the pump and of the first Stokes component, which specified the inhomogeneity in Eqs. (9).

It is easily seen that the anisotropy of the angular distributions is determined by the right-hand sides of Eqs. (9). We can therefore put for simplicity $p^2 = k_1^2$, $q^2 = k_{-2}^2$, and neglect the isotropic damping $\mathscr{E}_1$ due to the interaction with the pump, and the amplification $\mathscr{E}_{-2}$ due to the interaction with the field $\mathscr{E}_{-1}$. Since it is assumed that $|\mathbf{E}_{-2}| \ll |\mathbf{E}_0|$ and $|\mathbf{E}_1| \ll |\mathbf{E}_{-1}|$, the system (9) can be solved by considering the second terms in the right-hand sides as a perturbation. Then, at large distances from the active volume, we obtain for the amplitudes $\mathscr{E}_1$ and $\mathscr{E}_{-2}$:

$$\mathscr{E}_1 = \mathscr{E}_1^{(0)} + \mathscr{E}_1^{(1)}, \quad \mathscr{E}_{-2} = \mathscr{E}_{-2}^{(0)} + \mathscr{E}_{-2}^{(1)}$$

where

$$\mathscr{E}_1^{(0)}(r, \theta) = \frac{1}{2i}\chi_1 r_0 a_0^2 a_{-1}^* \frac{e^{-ik_1 r}}{r} \frac{e^{i\Delta k_1 l} - 1}{\Delta k_1} \frac{J_1(k_1 r_0 \sin\theta)}{k_1 \sin\theta}, \tag{11}$$

$$\mathscr{E}_{-2}^{(0)}(r, \theta) = \frac{1}{2i}\chi_{-2} r_0 a_0^* a_{-1}^2 \frac{e^{-ik_{-2}r}}{r} \frac{e^{i\Delta k_{-2}l} - 1}{\Delta k_{-2}} \frac{J_1(k_{-2} r_0 \sin\theta)}{k_{-2} \sin\theta}, \tag{11a}$$

where θ is the angle of the vector of the point of observation $\mathbf{r}$ with the z axis; $J_1(x)$ is a Bessel function of first order, $k_0 = |\mathbf{k}_0|$, $k_{-1} = |\mathbf{k}_{-1}|$,

$$\mathscr{E}_1^{(1)}(r, \theta) = \frac{r_0\chi_1^2\omega_{-2}^2}{16\pi^3\omega_1^2} a_0^2 |a_{-1}|^2 a_{-1} \frac{\exp\left(-ik_1 r\right)}{r} \int \left[\frac{\exp\left(i\Delta k_{-2}^{(\sigma)}l\right) - 1}{\Delta k_{-2}^{(\sigma)}} \frac{J_1(k_{-2}r_0 n_\perp^{(\sigma)})}{n_\perp^{(\sigma)}}\right.$$
$$\times \int_V \exp\left\{i\left(\mathbf{k}_0 + \mathbf{k}_{-1} - k_1\mathbf{n} - k_{-2}\mathbf{n}^{(\sigma)}\right)\mathbf{r}'\right\} dr'\bigg] d\sigma, \tag{12}$$

$$\mathscr{E}_{-2}^{(1)}(r, \theta) = -\frac{r_0\chi_{-2}^2\omega_1^2}{16\pi^3\omega_{-2}^2} |a_0|^2 a_0 a_{-1}^2 \frac{\exp\left(-ik_2 r\right)}{r} \int \left[\frac{\exp\left(-i\Delta k_1^{(\sigma)}l\right) - 1}{\Delta k_1^{(\sigma)}} \frac{J_1(k_1 r_0 n^{(\sigma)})}{n^{(\sigma)}}\right.$$
$$\times \int_V \exp\left\{-i\left(\mathbf{k}_0 + \mathbf{k}_{-1} - k_{-2}\mathbf{n} - k_1\mathbf{n}^{(\sigma)}\right)\mathbf{r}'\right\} dr'\bigg] d\sigma, \tag{12a}$$

$$\Delta k_1^{(\sigma)} = k_{-1} + k_1 n_\parallel^{(\sigma)} - 2k_0, \qquad \Delta k_{-2}^{(\sigma)} = k_0 + k_{-2} n_\parallel^{(\sigma)} - 2k_{-1}.$$

where $\int d\sigma$ denotes integration over the unit sphere, $\mathbf{n}^{(\sigma)}$ is a unit radius vector, $\mathbf{n}_{\parallel}^{(\sigma)}$ is the projection of $\mathbf{n}^{(\sigma)}$ on the z axis, $\mathbf{n}_{\perp}^{(\sigma)}$ is the projection of $\mathbf{n}^{(\sigma)}$ on the plane perpendicular to the z axis, and $\mathbf{n} = \mathbf{r}/|\mathbf{r}|$ is a unit vector in the $\mathbf{r}$ direction.

Thus, the fields $\mathbf{E}_1$ and $\mathbf{E}_{-2}$ constitute spherical waves with amplitudes that depend on the propagation direction $\mathbf{r}$. The terms $\mathcal{E}_1^{(0)}$ and $\mathcal{E}_{-2}^{(0)}$ are due to the contribution of a special radiation of the Cerenkov type, which results from the interaction of the parallel intense fields of the pump and of the first Stokes component, propagating with superluminal velocity through the polarization wave at the frequency of the considered component. We indicate that the possibility of realizing radiation of the Cerenkov type in the scattering was noted in [12, 13]. From (11) and (11a) it is easy to find that this radiation has, for each of the components,† first an axial maximum $\theta = 0$, and second, an off-axis maximum, which in the case of the first anti-Stokes component is directed at an angle determined from the relation

$$\cos \theta_1 = \frac{2k_0 - k_{-1}}{k_1},\tag{13}$$

and in the case of the second Stokes component from the condition

$$\cos \theta_{-2} = \frac{2k_{-1} - k_0}{k_{-2}}.\tag{13a}$$

As seen from (9), the terms $\mathcal{E}_{+1}^{(1)}$ and $\mathcal{E}_{-2}^{(1)}$ are due to the interactions of the components $\mathcal{E}_1$ and $\mathcal{E}_{-2}$ with each other and with the pump field and the first Stokes component field. It is easily seen that the integrals in (12) have a sharp maxima for definite off-axis directions of the vector $\mathbf{r}$. It can be shown that the directions to these maxima are determined by the vectors $\mathbf{k}_1$ and $\mathbf{k}_{-2}$ in accordance with the rule

$$\mathbf{k}_0 + \mathbf{k}_{-1} = \mathbf{k}_1 + \mathbf{k}_{-2},\tag{14}$$

where $|\mathbf{k}_1| = k_1$, $|\mathbf{k}_{-2}| = k_{-2}$. From (14) we easily obtain expressions for the angles between the vectors $\mathbf{k}_1$ and $\mathbf{k}_{-2}$ and the z axis:

$$\sin \theta_1 = \frac{k_{-2}}{k_1} \sin \theta_{-2},\tag{15}$$

$$\cos \theta_{-2} = \frac{(k_0 + k_{-1})^2 - (k_1^2 - k_{-2}^2)}{2k_{-2}(k_0 + k_{-1})}.\tag{15a}$$

The relations (13), (13a), (15), and (15a) provide a complete picture of the arrangement of the off-axis maxima. We note that the rule (14) for the description of radiation of class II was first proposed from phenomenological considerations concerning four-photon processes in [7].

Before we proceed to a comparison with experiment, we note that the positions of the maxima do not depend on the transverse inhomogeneity of the beam. In fact, if we specify the input fields in the form of Gaussian beams

$$\varepsilon_0 = a_0 \exp(-i\mathbf{k}_0 \mathbf{r}) \exp\left(-\frac{r_\perp^2}{2b_0^2}\right), \qquad \varepsilon_1 = a_{-1} \exp(-i\mathbf{k}_1 \mathbf{r}) \exp\left(-\frac{r_\perp^2}{2b_{-1}^2}\right),$$

we obtain equations that generalize (11) and (11a) to the case of inhomogeneous input fields:

$$\varepsilon_1^{(0)}(r, \theta) = \frac{\chi_1 a_0^2 a_{-1}^*}{2i} \frac{\exp(-ik_1 r)}{r} \frac{\exp(i\Delta k_1 l) - 1}{\Delta k_1} \int_0^\delta J_0(k_1 r_\perp \sin \theta_1) \exp\left[-r_\perp^2\left(\frac{1}{b_0^2} + \frac{1}{2b_{-1}^2}\right)\right] r_\perp dr_\perp,$$

$$\mathcal{E}_{-2}^{(0)}(r, \theta) = \frac{\chi_{-2} a_0^* a_{-1}^2}{2i} \frac{\exp(-ik_{-2} r)}{r} \frac{\exp(i\Delta k_{-2} l) - 1}{\Delta k_{-2}} \int_0^\delta J_0(k_{-2} r_\perp \sin \theta_{-2}) \exp\left[-r_\perp^2\left(\frac{1}{2b_0^2} + \frac{1}{b_{-1}^2}\right)\right] r_\perp dr_\perp,$$

where δ is the radius of the collimator ($\delta = +\infty$ in the case of unbounded beams). This leads at $b_0 \gg \delta$, $b_{-1} \gg \delta$ to Eqs. (11) and (11a). The obtained expressions show that for real Gaussian beams, at any δ, the positions of the maxima of the radiation are determined as before by Eqs. (13) and (13a). The inhomogeneity of the beam influences only the values of the intensities at these maxima. In particular, for unbounded beams

---

†After completion of [9] (see [8] for a preliminary publication), a paper [14] was published dealing with SRS in a focused beam, where an analogous result was obtained (in the limit of an infinitely elongated caustic).

TABLE 1. Off-Axis Radiation Angles (in **radians**)

Medium	Angles	Calculation by formulas		Experiment		
		(13), (13a)	(15), (15a)	[7]	[4]	[5]
Benzene	θ_1	0,0377	0,0367	0,035	0,0325	0,0313
	θ_{-2}	0,0404	0,0431	0,0385	0,038	—
Carbon disulfide	θ_1	0,0374	0,0372	—	0,033	—
	θ_{-2}	0,0393	0,0412	0,030	—	—
Nitrobenzene	θ_1	0,0649	0,0571	—	0,058	0,0569
	θ_{-2}	0,0690	0,0765	0,073	0,065	—

$$\mathscr{E}_1^{(0)}(r,\theta) = \frac{\chi_1 a_0^2 a_{-1}^*}{2i} \frac{\exp(-ik_1 r)}{r} \frac{\exp(i\Delta k_1 l)-1}{\Delta k_1} \frac{\exp\left[-k_1^2 \sin^2\theta_1/4\left(\frac{1}{b_0^2}+\frac{1}{2b_{-1}^2}\right)\right]}{2\left(\frac{1}{b_0^2}+\frac{1}{2b_{-1}^2}\right)},$$

$$\mathscr{E}_{-2}^{(0)}(r,\theta) = \frac{\chi_{-2} a_0^* a_{-1}^2}{2i} \frac{\exp(-ik_{-2}r)}{r} \frac{\exp(i\Delta k_{-2}l)-1}{\Delta k_{-2}} \frac{\exp\left[-k_{-2}^2 \sin^2\theta_{-2}/4\left(\frac{1}{2b_0^2}+\frac{1}{b_{-1}^2}\right)\right]}{2\left(\frac{1}{2b_0^2}+\frac{1}{b_{-1}^2}\right)}.$$

Table 1 lists the angles of the off-axis radiation, calculated from formulas (13), (13a), (15), and (15a) with account taken of the refraction by the boundary between the active volume and the air, and the experimental data of [7]. In addition, the last two columns show for comparison the experimental values of the angles θ_1 and θ_{-2} for class-II radiation [4, 5].

It is seen from the table that the angles of both maxima, for each of the components, are in good agreement with the experimental values [7] with account taken of the experimental error. We note that both the maxima with respect to θ_1 and the maxima with respect to θ_{-2} are separated by distances (on the average ~0.005 rad) that do not exceed the experimental errors (15% according to [7]). We can therefore conclude that the two-ring structure of the maxima of the second Stokes and first anti-Stokes components were not observed in the experiment because of the insufficient resolution.

Of particular interest is the good agreement with experiment for the angles between the maximum of the Cerenkov-type radiation (13) and (13a), which makes the main contribution by virtue of (11) and (12).

Thus, the results show that the presented analysis describes well the experimental data of [7]. In addition (see the table), the angles calculated from (13) and (15) are close to the experimental angles of class II [4, 5]. We can therefore draw a conclusion that agrees with the results of [7], namely that the primary role in the mechanism of the formation of rings II is played by the paraxial radiation of the first Stokes component. Under conditions when the paraxial part of this radiation is for some reason intense enough, radiation of class II should prevail over that of class I. We note in this connection that in liquids this cause may be the self-focusing effect. In conclusion it must be stated that further more detailed experiments in accord with the two-cell scheme are needed for a final answer to the question of the angular distributions of the SRS.

3. HIGHER STOKES COMPONENTS

In the preceding section, using as an example the second Stokes and the first anti-Stokes components, we have shown that if for some reason the paraxial part of the radiation of the first Stokes component is amplified in the active volume, the result is an angular distributions of the Cerenkov type. A similar situation was simulated by specifying the pump field and the field of the first Stokes component in the form of intense plane waves propagating strictly along the sample axis.

We consider below the angular distributions of the higher Stokes fields on the basis of the mechanism of consecutive generation of the components. It should be noted that the angular distributions of the higher components can be complex and varied, and the number of radiations of various types increases rapidly with the number of the component. We consider next some of the cases that are realistic from the experimental point of view and in which one of the possible types of radiation predominates. We assume the fields of the pump and of the first Stokes component to be given and to have constant amplitude. We note that the presence of a given field of the first Stokes component makes it unnecessary to introduce into the theory pointlike sources of their radiation at the first Stokes frequency. Polarization at the frequency of the m-th SRS component (m < 0 and

$m > 0$ correspond to Stokes and anti-Stokes component, respectively, while $m = 0$ corresponds to the pump) in the quasistatic scattering regime takes the form (6).

We consider next fields in narrow cones near the axis of the active volume, using the scalar form of Maxwell's equations

$$-(\Delta + k_m^2)\mathscr{E}_m = \mu_0\omega_m^2 P_m,$$

$$P_m = \frac{\chi}{\delta - i}\mathscr{E}_{m-1}\sum_{l=-\infty}^{\infty}\mathscr{E}_l\mathscr{E}_{l-1}^* + \frac{\chi}{\delta + i}\mathscr{E}_{m+1}\sum_{l=-\infty}^{\infty}\mathscr{E}_l^*\mathscr{E}_{l-1}, \qquad k_m^2 = \frac{\omega_m^2}{c_m^2}. \tag{16}$$

In accordance with the consecutive mechanism of excitation of the SRS components, we shall assume that

$$|\mathscr{E}_{l+1}| \ll |\mathscr{E}_l|, \qquad l > 0,$$
$$|\mathscr{E}_{l-1}| \ll |\mathscr{E}_l|, \qquad l < 0. \tag{17}$$

If at the same time we can neglect the depletion of the pump and of the first Stokes component, then we have for any value of l,

$$|\mathscr{E}_l| \ll |\mathscr{E}_{-1}| < |\mathscr{E}_0|. \tag{18}$$

We consider the field of the third Stokes component $\mathscr{E}_{-3}$. The fields of the pump and of the first Stokes components are assumed to be plane waves propagating along the axis of the active volume V (cylinder of length l and radius r_0)

$$\mathscr{E}_0 = a_0\exp(-i\mathbf{k}_0\mathbf{r}), \qquad \mathscr{E}_{-1} = a_{-1}\exp(-i\mathbf{k}_{-1}\mathbf{r}), \tag{19}$$

where $\mathbf{k}_0$ and $\mathbf{k}_{-1}$ are the wave vectors, with $\mathbf{k}_0 \| \mathbf{k}_{-1}$. According to (17), the main contributions to the polarization at the second and third Stokes frequencies are

$$P_{-2} = \frac{\chi}{\delta + i}(\mathscr{E}_0^*\mathscr{E}_{-1}^2 + |\mathscr{E}_{-1}|^2\mathscr{E}_{-2}), \tag{20}$$

$$P_{-3} = \frac{\chi}{\delta + i}(\mathscr{E}_{-1}^*\mathscr{E}_{-2}^2 + \mathscr{E}_0^*\mathscr{E}_{-1}\mathscr{E}_{-2} + |\mathscr{E}_{-2}|^2\mathscr{E}_{-3}). \tag{21}$$

The last terms in (20) and (21) correspond to two-photon processes of direct generation of a given component from the preceding one (which acts as the pump), and lead to its exponential growth within the active volume. During the early stage of generation of the second and third Stokes components, the conditions (18) are satisfied and the indicated terms in (20) and (21) can be neglected. Then, according to (16) and (20),

$$\mathscr{E}_{-2}(\mathbf{r}) = \frac{A_{-2}}{4\pi}\int_{(V)}\frac{\exp[-ik_{-2}|\mathbf{r}-\mathbf{r}'|]}{|\mathbf{r}-\mathbf{r}'|}\exp[i(\mathbf{k}_0 - 2\mathbf{k}_{-1})\mathbf{r}']\,d\mathbf{r}',$$

$$A_{-2} = \frac{\chi\mu_0\omega_{-2}^2}{\delta + i}a_0^*a_{-1}^2. \tag{22}$$

Neglecting by virtue of (18) the first terms in P_{-3} and using (22), we obtain at large distances from the active volume

$$\mathscr{E}_{-3}(r,\theta) = \frac{A_{-2}A_{-3}r_0}{16\pi^2}\frac{e^{-ik_{-3}r}}{r}\int d\sigma\left[\frac{\exp(i\Delta k_{-2}l) - 1}{\Delta k_{-2}}\frac{J_1(r,k_{-2}|\mathbf{n}_\perp^{(\sigma)}|)}{|\mathbf{n}_\perp^{(\sigma)}|}\int_{(V)}\exp\{i(\mathbf{k}_0 + k_{-3}\mathbf{n} - \mathbf{k}_{-1} - k_{-2}\mathbf{n}^{(\sigma)})\mathbf{r}'\}\,d\mathbf{r}'\right],$$

$$\Delta k_{-2} = k_0 - 2k_{-1} + k_{-2}n_\parallel^{(\sigma)}, \qquad A_{-3} = \frac{\chi\mu_0\omega_{-3}^2}{\delta + i}a_0^*a_{-1}. \tag{23}$$

where θ is the angle of the vector r with the z axis. It is easily seen that (23) has a sharp maximum for the direction n defined by the vector $\mathbf{k}_{-3}$ in accordance with the rule

$$\mathbf{k}_0 - \mathbf{k}_{-1} = \mathbf{k}_{-2} - \mathbf{k}_{-3}, \tag{24}$$

where $|\mathbf{k}_{-3}| = k_{-3}$, $|\mathbf{k}_{-2}| = k_{-2}$. The scattered radiation is directed mainly along the generators of a cone that makes with the z axis an angle

$$\theta_{-3}^{(1)} = \arccos\frac{k_{-2}^2 - k_{-3}^2 - (k_0 - k_{-1})^2}{2k_{-3}(k_0 - k_{-1})}. \tag{25}$$

We consider now the case when the decisive role in the formation of the angular distribution of the field $\mathscr{E}_{-3}$ is played by the first term in the polarization (21). This is possible if the pump has been depleted, and the

128

intensity of the second Stokes component has strongly increased on account of the direct process of generation of $\mathcal{E}_{-2}$ from the first Stokes component. In this case, for a strongly elongated active volume ($r_0/l \ll 1$) the radiation $\mathcal{E}_{-2}$ is mainly directed along the z axis. We assume for simplicity that besides the first Stokes field there is given also the field of the second Stokes component in the form of an intense plane wave of constant amplitude and propagating along the z axis:

$$\mathcal{E}_{-2} = a_{-2} \exp(-i\mathbf{k}_2\mathbf{r}).$$

Then

$$\mathcal{E}_{-3}(r, \theta) = \frac{\chi\mu_0 r_0 \omega_{-3}^2}{2i(\delta + i)} a_{-1}^* a_{-2}^2 \frac{e^{-ik_{-3}r}}{r} \frac{e^{i\Delta k_{-3}l} - 1}{\Delta k_{-3}} \frac{J_1(k_{-3}r_0 \sin\theta)}{k_{-3}\sin\theta},$$
$$\Delta k_{-3} = k_{-1} + k_{-3}\cos\theta - 2k_{-2}. \tag{26}$$

According to (26), the emission of the third Stokes component has an axial maximum ($\theta = 0$) and an off-axis maximum. The off-axis radiation is directed at an angle determined from the relation

$$\cos\theta_{-3}^{(2)} = \frac{2k_{-2} - k_{-1}}{k_{-3}}. \tag{27}$$

The radiation at the angle $\theta_{-3}^{(2)}$ is of the Cerenkov type.

Similar arguments can be used also for the fourth Stokes component. The principal terms of the polarization P_{-4} are of the form

$$P_{-4} = \frac{\chi}{\delta + i}(\mathcal{E}_{-2}^* \mathcal{E}_{-3}^2 + \mathcal{E}_0^* \mathcal{E}_{-1}\mathcal{E}_{-3} + \mathcal{E}_{-1}^* \mathcal{E}_{-2}\mathcal{E}_{-3} - |\mathcal{E}_{-3}|^2\mathcal{E}_{-1}). \tag{28}$$

In the case of depletion of the pump and of the first Stokes component simultaneously with a strong growth of the third component via the direct process $\mathcal{E}_{-3} \to \mathcal{E}_{-4}$ the polarization (28) is determined by the first term: $P_{-4} \sim \mathcal{E}_{-2}^* \mathcal{E}_{-3}^2$. Then, in analogy with (26) and (27) we can find that the emission $\mathcal{E}_{-4}(r, \theta)$ has an axial maximum $\theta = 0$ and an off-axis maximum (Cerenkov radiation) at an angle

$$\theta_{-4} = \arccos\frac{2k_{-3} - k_{-2}}{k_{-4}}. \tag{29}$$

By considering a similar sequential generation mechanism for the m-th Stokes component we obtain at large distances from the active object (under the condition $|\mathcal{E}_{-m+3}| \ll |\mathcal{E}_{-m+1}|$)

$$\mathcal{E}_{-m}(r, \theta) = \frac{\chi\mu_0 r_0 \omega_{-m}^2}{2i(\delta + i)} a_{-m+2}^* a_{-m+1}^2 \frac{e^{-ik_{-m}r}}{r} \frac{e^{i\Delta k_{-m}l} - 1}{\Delta k_{-m}} \frac{J_1(k_{-m}r_0 \sin\theta)}{k_{-m}\sin\theta},$$
$$\Delta k_{-m} = k_{-m+2} + k_{-m}\cos\theta - 2k_{-m+1}. \tag{30}$$

The direction to the off-axis maximum is determined by the rule

$$\theta_{-m} = \arccos\frac{2k_{-m+1} - k_{-m+2}}{k_{-m}}. \tag{31}$$

Table 2 lists the values of the angles (27) and (29) of the off-axis maxima with allowance for refraction by the boundary between the active volume and the air as well as the experimental data of [7]. It follows from the table that the radiation cones (31) were observed in the experiment [7].

The results show that in the presence of intense axial radiation of the first Stokes component, the mechanism of the development of the higher Stokes component differs from the Townes mechanism. The radiations produced under these conditions evolve either on account of various four-wave processes of the type (24) or on account of generation of Cerenkov-type radiation.

4. SRS IN CROSSED PUMP AND FIRST-STOKES-COMPONENT FIELDS

The results of Sections 2 and 3 show that, depending on the concrete conditions of the experiment, the SRS radiation can vary greatly in character. Interest attaches therefore to the possibility of realizing the various types of SRS radiation as functions of these conditions. We consider first the radiation of the SRS components, assuming the pump field to be a plane wave with constant amplitude:

$$\mathbf{\varepsilon}_0 = \mathbf{a}_0 \exp(-i\mathbf{k}_0\mathbf{r}), \tag{32}$$

which occupies in the medium a cylindrical volume V of length l and radius r_0 (the cylinder axis is z). We assume that the input beam of the first Stokes component (plane wave) is directed at an angle ψ to the z axis and

TABLE 2. Off-Axis Radiation Angles of the Third and Fourth Stokes Components (in **radians**)

Medium	Angles	Calculation by formulas (27) and (29)	Experiment [7]
Benzene	θ_{-3}	0.041	0.036; 0.037
	θ_{-4}	0.0375	0.035; 0.037
Carbon disulfide	θ_{-3}	0.0385	0.034
	θ_{-4}	0.0382	0.033
Nitrobenzene	θ_{-3}	0.0663	0.072

the region of intersection of the beams overlaps the entire volume V. In this formulation of the problem, which corresponds to experiments by the two-cell scheme, the intensity of the Stokes wave greatly exceeds the intensity of any other component. It can therefore also be regarded as a given wave with constant amplitude

$$\varepsilon_{-1}(\mathbf{r}, t) = \mathbf{a}_{-1} \exp(-i\mathbf{k}_{-1}\mathbf{r}),$$

where $\cos \psi = \mathbf{k}_0 \mathbf{k}_{-1} / k_0 k_{-1}$. Using the inequalities $|\mathbf{E}_{l+1}| \ll |\mathbf{E}_l|$, $l > 0$; $|\mathbf{E}_{l+1}| \ll |\mathbf{E}_l|$, $l < 0$, and writing down Maxwell's equations in the scalar approximation, we have for the second Stokes and first anti-Stokes components

$$(\Delta + k_1^2)\mathscr{E}_1 = -\chi_1[a_0^2 a_{-1}^* \exp\{-i(2\mathbf{k}_0 - \mathbf{k}_{-1})\mathbf{r}\} + a_0 a_{-1}\mathscr{E}_{-2}^* \exp\{-i(\mathbf{k}_0 + \mathbf{k}_{-1})\mathbf{r}\}]g(\mathbf{r}),$$

$$(\Delta + k_{-2}^2)\mathscr{E}_{-2} = -\chi_{-2}[a_0 a_{-1}^2 \exp\{-i(2\mathbf{k}_{-1} - \mathbf{k}_0)\mathbf{r}\} + a_0 a_{-1}\mathscr{E}_1^* \exp\{-i(\mathbf{k}_0 + \mathbf{k}_{-1})\mathbf{r}\}]g(\mathbf{r}). \tag{33}$$

The principal contributions to the amplitudes at large distances from the active volume are

$$\mathscr{E}_1^{(0)}(\mathbf{r}) = \frac{1}{2}\chi_1 r_0 a_0^2 a_{-1}^* \frac{\exp(-ik_1 r)}{r} f_1(\Delta k_1^{\parallel}) f_2(\Delta k_1^{\perp}), \tag{34}$$

$$\mathscr{E}_{-2}^{(0)}(\mathbf{r}) = \frac{1}{2}\chi_{-2} r_0 a_0^2 a_{-1}^* \frac{\exp(-ik_{-2} r)}{r} f_1(\Delta k_{-2}^{\parallel}) f_2(\Delta k_{-2}^{\perp}), \tag{35}$$

where the functions f_1 and f_2 are defined by the relations

$$f_1(\xi) = \frac{\exp(i\xi l) - 1}{\xi}, \qquad f_2(\xi) = \frac{J_1(\xi r_0)}{\xi}; \tag{36}$$

$$\Delta k_1^{\parallel} = k_{-1}\cos\psi + k_1\cos\theta - 2k_0, \quad \mathbf{n} = \mathbf{r}/|\mathbf{r}|,$$

$$\Delta k_1^{\perp} = |\mathbf{k}_{-1}^{\perp} - k_1\mathbf{n}^{\perp}| = [(k_1\sin\theta\sin\varphi)^2 + (k_{-1}\sin\psi - k_1\sin\theta\cos\varphi)^2]^{1/2},$$

$$\Delta k_{-2}^{\parallel} = k_0 + k_{-2}\cos\theta - 2k_{-1}\cos\psi, \tag{37}$$

$$\Delta k_{-2}^{\perp} = |k_{-2}\mathbf{n}^{\perp} - 2\mathbf{k}_{-1}^{\perp}| = [(k_{-2}\sin\theta\sin\varphi)^2 + (2k_{-1}\sin\psi - k_{-2}\sin\theta\cos\varphi)^2]^{1/2},$$

$\mathbf{k}_{-1}^{\perp}$, $\mathbf{n}^{\perp}$ are the projections of the corresponding vectors on a plane perpendicular to the z axis; θ is the angle between the vector $\mathbf{n}$ and the z axis, and φ is the angle between the plane of the vectors $\mathbf{k}_0$, $\mathbf{k}_{-1}$ and the plane of the vectors $\mathbf{k}_0$, $\mathbf{r}$ (Fig. 1).

It is easily seen that the maxima of the intensities of both components are located at $\varphi = 0$, i.e., in the plane of the vectors $\mathbf{k}_0$ and $\mathbf{k}_{-1}$. The character of the dependences of these maxima on the angle θ is determined by the value of the crossing angle ψ. We consider first the anti-Stokes field.

If ψ is equal to the Townes value†

$$\psi = \psi_1^T = \arccos\frac{k_{-1}^2 + 4k_0^2 - k_1^2}{4k_0 k_{-1}},$$

then (5) has a sharp maximum at

$$\theta_1^{(1)} = \arccos\frac{k_1^2 + 4k_0^2 - k_{-1}^2}{4k_0 k_1}. \tag{38}$$

At this value of the angle θ the quantities $\Delta k_1^{\parallel}$ and $\Delta k_1^{\perp}$ are equal to zero and both functions f_1 and f_2 reach their maximum values $f_1(0) = il$, $f_2(0) = r_0/2$. The direction of the maximum radiation is determined thus in accord with the triangle rule $\mathbf{k}_{-1} + \mathbf{k}_1 = 2\mathbf{k}_0$. The half-width of the Townes maximum with respect to the angle

---

†Here and below we present angle values calculated for a surrounding medium having the same refractive indices as the investigated substance.

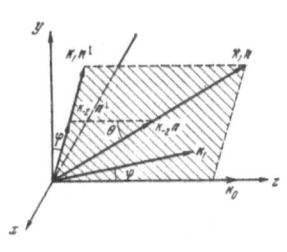

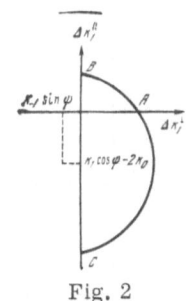

Fig. 1   Fig. 2

θ is †

$$\Delta\bar{\theta}_1^{(1)} = \min\left\{\frac{2\pi}{k_1\theta_1^{(1)}l};\ \frac{2.73}{k_1r_0}\right\}. \tag{39}$$

Its half-width with respect to the angle φ is

$$\Delta\bar{\varphi}_1^{(1)} = 2.73/k_1r_0\theta_1^{(1)}. $$

Thus, the Townes radiation on a screen placed perpendicular to the z axis produces the spot with horizontal dimension $2\Delta\bar{\varphi}_1^{(1)}$ and with vertical dimension $2\Delta\bar{\theta}_1^{(1)}$.

If $\psi \neq \psi_1^T$, the condition $\mathbf{k}_{-1} + \mathbf{k}_1 = 2\mathbf{k}_0$ cannot be satisfied. In this case, according to (34) and (35), it is necessary to test for the maximum the functions $f(\Delta k_1^{\parallel}, \Delta k_1^{\perp}) = f_1(\Delta k_1^{\parallel})f_2(\Delta k_1^{\perp})$ of the two variables $\Delta k_1^{\parallel}$ and $\Delta k_1^{\perp}$, which are connected by the relation

$$[\Delta k_1^{\parallel} - (k_{-1}\cos\psi - 2k_0)]^2 + [\Delta k_1^{\perp} + k_{-1}\sin\psi]^2 = k_1^2. \tag{40}$$

In the general case this maximum lies at one of the points on the arc BAC of the circle (40) (Fig. 2). The point A with coordinates $\Delta k_1^{\parallel} = 0$, $\Delta k_1^{\perp} = -k_{-1}\sin\psi + \sqrt{k_1^2 - (2k_0 - k_{-1}\cos\psi)^2}$ corresponds to the maximum of the function f_1; the corresponding value of the angle θ is

$$\theta_1^{(2)} = \arccos\frac{2k_0 - k_{-1}\cos\psi}{k_1}. \tag{41}$$

The point B with coordinates $\Delta k_1^{\parallel} = \sqrt{k_1^2 - k_{-1}^2\sin^2\psi} + k_{-1}\cos\psi - 2k_0$, $\Delta k_1^{\perp} = 0$ corresponds to the maximum of the function f_2; the corresponding value of the angle θ is

$$\theta_1^{(3)} = \arccos\sqrt{1 - \left(\frac{k_{-1}}{k_1}\sin\psi\right)^2}. \tag{42}$$

The real values of the parameters k_0, k_1, k_{-1} and r_0, l are such that

$$\Delta k_1^{\parallel}(\theta_1^{(3)})\,l \gg 1, \qquad \Delta k_1^{\perp}(\theta_1^{(2)})\,r_0 \gg l. \tag{43}$$

Therefore the function f_1, which has a maximum at the point A, oscillates rapidly for displacements along the arc ABC away from this point; its envelope attenuates rapidly. The function f_2 behaves similarly on moving from the point B to the point A. It is clear therefore that the function f has two maxima in the vicinity of the points A and B. We consider in greater detail the structure of these maxima, which results from the diffraction of the scattered radiation by the boundaries of the active medium.

In the vicinity of the point A we have

$$\Delta k_1^{\parallel}(\theta) = -k_1\theta_1^{(2)}\Delta\theta, \qquad \Delta k_1^{\perp}(\theta) = (k_1\theta_1^{(2)} - k_{-1}\psi) + k_1\Delta\theta. \qquad |\Delta\theta| \ll \theta_1^{(2)} \tag{44}$$

and the oscillations of the function f_1 have a period, which characterizes also the rate of damping of this function, equal to $\Delta\bar{\theta}_1^{(2)} = 2\pi/k_1\theta_1^{(2)}l$. The period of the oscillations of the function f_2 is $\Delta\bar{\bar{\theta}}_1^{(2)} = 2\pi/k_1r_0$ and, according to (44), the envelope of the function varies weakly in this case. Therefore the half-width of the maximum is determined by the value of $\Delta\bar{\theta}_1^{(2)}$, while the number of oscillations f at the half-width is $m_1^{(2)} = \Delta\bar{\theta}_1^{(2)}/\Delta\bar{\bar{\theta}}_1^{(2)} = r_0/\theta_1^{(2)}l$. The value of the angle $\theta_1^{(2)}$ determines the symmetry axis of the diffraction structure with respect to angle θ. The value of the maximum point of the function f lies in the range from $\theta_1^{(2)}$ to $\theta_1^{(2)} + \Delta\bar{\theta}_1^{(2)}$ and can only

---

†We assume for the half-width the largest of the values of the angle θ at which $f_1(\Delta k_1^{\parallel})$ vanishes for the first time, and of the value of θ at which $|f_2(\Delta k_1^{\perp})|^2$ decreases to one-tenth of its maximum value.

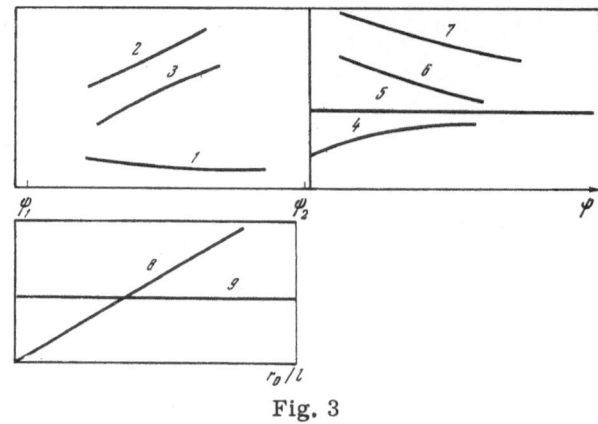

Fig. 3

accidentally be located at the point $\theta_1^{(2)}$. In view of the rapid oscillations of f_2, the diffraction structure (the disposition of the local maxima of the structure and the values of f at these points) is very sensitive to small changes of the parameters r_0 and ψ.

We now take into account the fact that the diffraction divergence Δ of the real input beams ($\sim 2\pi/k_0 r_0$ for the pump and $\sim 2\pi/k_{-1} r_0$ for the Stokes field) is a quantity of the same order as the characteristic dimension $\Delta\bar{\theta}_1^{(2)}$ of the diffraction pattern. This means that under real conditions the diffraction structure of the maximum (41) with respect to the angle θ is "smeared out." At the same time the total intensity of the maximum changes radically following small changes of r_0 and ψ. It should therefore change strongly, as if randomly, from experiment to experiment, and should sometimes not appear at all. Therefore the maxima of (41) and (42), which have similar properties (see below), will be called "flickering."

We consider now the diffraction structure of the radiation (41) relative to the angle φ. At a fixed $\theta = \theta_1^{(2)}$, the angle φ can vary from 0 to 2π; the vector $\mathbf{r}$ of the observation point, making an angle $\theta_1^{(2)}$ with $\mathbf{k}_0$, describes in this case a cone (a ring on the screen), whose axis coincides with the direction of the vector $\mathbf{k}_0$. Whereas for Townes radiation (38) deviation of φ from the zero value, i.e., departure of the vector $\mathbf{r}$ from the plane of the vectors $\mathbf{k}_0$ and $\mathbf{k}_{-1}$, causes the intensity to attenuate sharply, in the present case this may not occur. If $\psi = 0$, then according to (34) and (36) the radiation intensity does not depend on φ. This means that the flickering maximum of (41) produces on the screen a ring with half-width $\Delta\bar{\theta}_1^{(2)}$, and with intensity that varies in the same manner in each experiment in all the radial sections of the ring. If $\psi \neq 0$, a diffraction structure is produced: the intensity of the ring oscillates in complicated fashion over the circles that make up the ring; the intensity in the general case drops from the upper part of the ring ($\varphi \approx 0$) towards the lower one ($\varphi \approx \pi$). Let us examine the optimal conditions for the observation of this structure. According to (34), (36), and (41), the condition that the envelope of the intensity be constant at all points of the ring is

$$\psi \ll \psi_1 = \frac{1}{3\sqrt{2}} \sqrt{\frac{k_1(k_1 + k_{-1} - 2k_0)}{k_{-1}^2}}.$$

When this condition is satisfied, the intensity oscillates at $\psi \gg \psi_2 = 1/4k_{-1}r_0$. The period of the oscillations is a variable quantity that depends on the value of the angle φ. Its maximum value is $\Delta\bar{\bar{\varphi}}_{1\max}^{(2)} = \sqrt{2\pi/k_{-1}\psi r_0}$ in the upper and lower parts of the ring ($\varphi \sim 0, \pi$) and its minimal value $\Delta\bar{\bar{\varphi}}_{1\min} = \pi/k_{-1}\psi r_0$ at the central parts of the ring ($\varphi \sim \pi/2, 3\pi/2$). The number of maxima of the diffraction structure is of the order of $n_1^{(2)} \sim k_1\psi r_0$. The divergence of the input beams does not exert substantial influence on the diffraction structure relative to the angle φ.

Figure 3 shows plots of $\Delta\bar{\theta}_1^{(2)}$ (curve 4), $\Delta\bar{\bar{\varphi}}_{1\max}^{(2)}$ (curve 1), and $n_1^{(2)}$ (curve 2) against the crossing angle ψ. The dependence of the ratio of the radial and angular dimensions of the spots of the annular diffraction structure $\Delta\bar{\theta}_1^{(2)}/\Delta\bar{\bar{\varphi}}_{1\max}^{(2)}$ is shown by curve 3.

In the vicinity of the point B (Fig. 2) at $\varphi = 0$ we have

$$\Delta k_1^\parallel(\Delta\theta) = k_1 + k_{-1} - 2k_0 - \frac{k_{-1}\psi^2}{2} - \frac{k_1(\theta_1^{(3)})^2}{2} - k_1\theta_1^{(3)}\Delta\theta - k_1\frac{(\Delta\theta)^2}{2},$$

$$\Delta k_1^\perp(\Delta\theta) = k_1\Delta\theta.$$

The width of the maximum of the radiation in terms of the angle θ in the direction of $\theta_1^{(3)}$ is $\Delta\bar{\theta}_1^{(3)} = 2.73/k_1 r_0$ (Fig. 3, curve 5), the value of $\theta_1^{(3)}$ giving the position of the lower edge of the diffraction pattern. At $\psi \neq 0$ the period of the oscillations f_1 is $\Delta\bar{\bar{\theta}}_1^{(3)} = 2\pi/k_1\theta_1^{(3)}l$, and their number is

$$m_1^{(3)} = \frac{2.73}{2\pi}\frac{\theta_1^{(3)}l}{r_0} \quad \text{for} \quad \psi = 0: \Delta\bar{\bar{\theta}}_1^{(3)} = \sqrt{\frac{4\pi}{k_1 l}}, \quad m_1^{(3)} = \frac{2.73}{\sqrt{4\pi}}\sqrt{\frac{l}{k_1 r_0^2}}.$$

The total intensity of the maximum changes strongly with small changes of the length l and of the cross angle ψ. It should therefore vary greatly from experiment to experiment. When φ deviates from the zero value, the intensity of the radiation attenuates strongly. Therefore the maximum of (42) produces on the screen a spot with vertical width $\Delta\bar{\theta}_1^{(3)}$ and with horizontal half-width $\Delta\bar{\varphi}_1^{(3)} = 2.73/(k_{-1}\psi r_0)$ (Fig. 3, curve 6). The ratio of the widths of the maximum in the horizontal and vertical directions is determined only by the value of the angle ψ: $2\Delta\bar{\varphi}_1^{(3)}/\Delta\bar{\theta}_1^{(3)} = 2k_1/k_{-1}\psi$ (Fig. 3, curve 7).

The comparative characteristics of the maxima are the following: the ratio of the width of the ring (41) to the vertical width of the spot (42) is $\frac{\Delta\bar{\theta}_1^{(2)}}{\Delta\bar{\theta}_1^{(3)}} = \frac{4\pi}{2.73\theta_1^{(2)}}\frac{r_0}{l} \approx \frac{4\pi}{2.73}\sqrt{\frac{k_1}{2(k_1+k_{-1}-2k_0)}}\frac{r_0}{l}$, i.e., it is determined only by the degree of elongation of the sample (Fig. 3, curve 8); the ratio of the longitudinal width of the diffraction spots of the ring structure in the central part of the ring to the horizontal width of the spot (42) does not depend on the geometrical parameters and on the dispersion properties of the medium, $\Delta\bar{\varphi}_{1\,\text{min}}^{(2)}/\Delta\bar{\varphi}_1^{(3)} = \pi/5.46$ (curve 9 on Fig. 3). We note also that the intensity of radiation under the Townes angle at $\psi = \psi_1^T$ is always higher than the intensity of the radiation (41), (42) produced at $\psi \neq \psi_1^T$.

We present numerical data for benzene in the case $\psi = 0$: $\theta_1^{(2)} \simeq 0.025$, $\theta_1^{(3)} = 0$. The values of $\Delta k_1^\perp$ at the point A and $\Delta k_1^\parallel$ at the point B are, respectively, $\sim583 r_0$ (cm) and $\sim7.4l$ (cm), i.e., the conditions (43) are actually satisfied for the usual r_0 and l. The widths of the maxima are $2\Delta\bar{\theta}_1^{(2)} = 0.022/l$(cm), $\Delta\bar{\theta}_1^{(3)} = 0.00012/r_0$(cm). To observe the diffraction structure of the ring, the crossing angle ψ should lie in the range $2\cdot10^{-5}/r_0$(cm) $< \psi < 0.01$.

We note in conclusion that by virtue of (41) our analysis was applicable to the case of crossing angle bounded from above by the value $\psi_{\text{max}} = \arccos[(2k_0 - k_1)/k_{-1}]$. If $\psi > \psi_{\text{max}}$, then the intensity of the "flickering" maximum (41) decreases sharply and the picture of the spatial distribution may turn out to be strongly diffused.

Similar investigations of the field of the second Stokes component (35), (37) show that its intensity has maxima of the Townes type at the point

$$\theta_{-2}^{(1)} = \arccos\frac{4k_{-1}^2 - k_0^2 - k_{-2}^2}{2k_0 k_{-2}} \tag{45}$$

at a crossing angle

$$\psi = \psi_{-2}^T = \arccos\frac{4k_{-1}^2 + k_0^2 - k_{-2}^2}{4k_0 k_{-1}}. \tag{46}$$

The angles (45) and (46) are obtained from the triangle rule $2\mathbf{k}_{-1} = \mathbf{k}_0 + \mathbf{k}_{-2}$. The half-widths of the Townes maximum are equal to

$$\Delta\bar{\theta}_{-2}^{(1)} = \min\left\{\frac{2\pi}{k_{-2}\theta_{-2}^{(1)}l}, \frac{2.73}{k_{-2}r_0}\right\}, \qquad \Delta\bar{\varphi}_{-2}^{(1)} = \frac{2.73}{k_{-2}r_0\theta_{-2}^{(1)}}.$$

At $\psi \neq \psi_{-2}^T$ there are two "flickering" maxima in the direction of the angles

$$\theta_{-2}^{(2)} = \arccos\frac{2k_{-1}\cos\psi - k_0}{k_{-2}}, \tag{47}$$

$$\theta_{-2}^{(3)} = \arccos\sqrt{1 - \left(\frac{2k_{-1}}{k_{-2}}\sin\psi\right)^2}. \tag{48}$$

Their characteristic values are

$$\Delta\bar{\theta}_{-2}^{(2)} = \frac{2\pi}{k_{-2}\theta_{-2}^{(2)}l}, \qquad \Delta\bar{\bar{\theta}}_{-2}^{(2)} = \frac{2\pi}{k_{-2}r_0}, \qquad \Delta\bar{\varphi}_{-2}^{(2)}(\varphi \sim 0, \pi) = \sqrt{\frac{\pi}{k_{-1}\psi r_0}},$$

$$\Delta\bar{\varphi}_{-2}^{(2)}\left(\varphi \sim \frac{\pi}{2}, \frac{3}{2}\pi\right) = \frac{\pi}{2k_{-1}\psi r_0}, \qquad n_{-2}^{(2)} \sim 2k_{-1}\psi r_0,$$

$$\Delta\bar{\theta}_{-2}^{(3)} = \frac{2.73}{k_1 r_0}, \qquad \Delta\bar{\bar{\theta}}_{-2}^{(3)} = \frac{2\pi}{k_{-2}\theta_{-2}^{(3)}l}, \qquad \Delta\bar{\bar{\theta}}_{-2}^{(3)}(\psi = 0) = \sqrt{\frac{4\pi}{k_{-2}l}},$$

$$\Delta\bar{\varphi}_{-2}^{(3)}(\psi \neq 0) = \frac{2.73}{2k_{-1}\psi r_0}$$

(the values of $\Delta\overline{\overline{\varphi}}_{-2}^{(2)}$ and $n_{-2}^{(2)}$ are given for angles ψ satisfying the condition $(8k_{-1}r_0)^{-1} \ll \psi \ll \frac{1}{6\sqrt{2}}\left[\frac{k_{-2}(k_{-2}+k_0-2k_{-1})}{k_{-1}^2}\right]^{1/2}$).

We note that at $\psi = 0$ the values of the angles (41) and (47) correspond to Cerenkov radiation, and the values (42) and (48) to axial radiation.

In analogy with Section 2, we can calculate the small increments produced in the amplitude by the second terms in the right-hand sides of Eqs. (33):

$$\mathscr{E}_1^{(1)}(r) = \frac{i\chi_1\chi_{-2}^* a_0^2 |a_{-1}|^2 a_{-1}}{32\pi^3 r} e^{-ik_1 r} k_{-2} \int d\sigma \Big[\int_{(V)} \exp\{i(k_0 - 2k_{-1} + k_{-2}n^{(\sigma)}) r'\} dr'$$

$$\times \int_{(V)} \exp\{-i(k_0 + k_{-1} - k_{-2}n - k_1 n^{(\sigma)}) r''\} dr'' \Big], \tag{49}$$

$$\mathscr{E}_{-2}^{(1)}(r) = \frac{i\chi_1^*\chi_{-2} k_1 |a_0|^2 a_0 a_{-1}^2}{32\pi^3 r} e^{-ik_{-2}r} \int d\sigma \Big[\int_{(V)} \exp\{-i(2k_0 - k_{-1} - k_1 n^{(\sigma)}) r'\} dr'$$

$$\times \int_{(V)} \exp\{i(k_0 + k_{-1} - k_{-2}n - k_1 n^{(\sigma)}) r''\} dr'' \Big], \tag{50}$$

where $\int d\sigma$ denotes integration over the unit sphere; $n^{(\sigma)}$ is the unit radius vector.† It is easily seen that the amplitudes (49) and (50) have sharp maxima for definite off-axis directions of the vector r. The directions to these maxima are determined by the rule

$$\mathbf{k}_0 + \mathbf{k}_{-1} = \mathbf{k}_1 + \mathbf{k}_{-2}, \tag{51}$$

with $|\mathbf{k}_1| = k_1$, $|\mathbf{k}_{-2}| = k_{-2}$. It follows from (51) that the radiation due to the additional contributions to the amplitudes is directed along generators of cones whose axis make an angle

$$\vartheta = \text{arctg}\left(\frac{k_0 - k_{-1}}{k_0 + k_{-1}} \text{tg} \frac{\psi}{2}\right) + \frac{\psi}{2} \tag{52}$$

with the z axis. The aperture angles of the cones of the anti-Stokes radiation and of the second Stokes component are, respectively,

$$\vartheta_1 = 4\,\text{arctg}\left(\frac{r_2}{r_1 - k_{-2}}\right), \tag{53}$$

$$\vartheta_{-2} = 4\,\text{arctg}\left(\frac{r_2}{r_1 - k_1}\right), \tag{54}$$

where

$$r_1 = \frac{1}{2}\left(k_0 + k_{-2} + k_0 \frac{\sin\psi}{\sin\vartheta}\right), \qquad r_2 = \left[\frac{(r_1 - k_1)(r_1 - k_{-2})\left(r_1 - k_0 \frac{\sin\psi}{\sin\vartheta}\right)}{r_1}\right]^{1/2}.$$

The maxima (53) and (54), as well as the Townes radiation, are stable to changes of the geometric parameters.

Thus, at $\psi = \psi_1^T(\psi_{-2}^T)$ the bulk of the radiation of the anti-Stokes field (the field of the second Stokes component) makes an angle $\theta_1^1(\theta_{-2}^1)$ with the z axis and produces on the screen a spot with dimensions $\Delta\overline{\varphi}_1^{(1)}$, $\Delta\overline{\theta}_1^{(1)}$. At $\psi \neq \psi^T$ radiation of lower intensity is produced (the maxima of $\theta_1^{(2)}$ and $\theta_1^{(3)}$ for the anti-Stokes wave and of $\theta_2^{(2)}$, $\theta_{-2}^{(3)}$ for the second Stokes component), which produce on the screen rings with a resolvable diffraction substructure (provided the conditions stated above on the value of ψ are satisfied; if not, only the upper part of the ring may appear), and spots with dimension $\Delta\overline{\varphi}_{1,-2}^{(3)}$, $\Delta\overline{\theta}_{1,-1}^{(3)}$. Their intensity varies strongly and seemingly arbitrarily from experiment to experiment. In all cases there is additional radiation (53), (54), whose axis coincides with the vector $\mathbf{k}_0 + \mathbf{k}_{-1}$. This radiation produces on the screen rings of low intensity with center shifted away from the z axis.

5. AXIALLY SYMMETRICAL FIELD OF FIRST STOKES COMPONENT

Let the Stokes field be directed along the generators of the cone that makes an angle ψ with the axial pump beam, so that at each point of the cylindrical volume filled by the pump beam there is an aggregate of plane waves of the first Stokes component, whose wave vectors lie on a conical surface with apex angle 2ψ. The corresponding experiment can be carried out by passing an axial beam of the first Stokes component through a ring diaphragm and focusing with the aid of a long-focus lens onto the location of the cell.

---

†To obtain these expressions it is necessary to determine from (33) the fields inside the active volume and to use their expansions in plane waves.

In this case the pump field takes the form (32), and the Stokes field is given by

$$\mathscr{E}_{-1} = a_{-1} \exp\left(- i k_{-1} \cos \psi z\right) \int_0^{2\pi} \exp\left(- i \mathbf{k}_{-1}^{\perp} \mathbf{r}^{\perp}\right) d\gamma = 2\pi a_{-1} \exp\left(- i k_{-1} z \cos \psi\right) J_0\left(k_{-1} | \mathbf{r}^{\perp}| \sin \psi\right), \tag{55}$$

where $\mathbf{r}^{\perp}$ and $\mathbf{k}_{-1}^{\perp}$ are the projections of the corresponding vectors on a plane perpendicular to the z axis, and γ is the angle between these projections. The main contributions to the amplitude of the first anti-Stokes and second Stokes components are

$$\mathscr{E}_1^{(0)} (\mathbf{r}) = \frac{\chi_1}{4\pi} \frac{\exp\left(- i k_1 r\right)}{r} \int_{(V)} \mathscr{E}_0^2 (\mathbf{r}') \mathscr{E}_{-1}^* (\mathbf{r}') \exp\left(i k_1 \mathbf{n} \mathbf{r}'\right) d\mathbf{r}', \tag{56}$$

$$\mathscr{E}_{-2}^{(0)} (\mathbf{r}) = \frac{\chi_{-2}}{4\pi} \frac{\exp\left(- i k_{-2} r\right)}{r} \int_{(V)} \mathscr{E}_0^* (\mathbf{r}') \mathscr{E}_{-1}^2 (\mathbf{r}') \exp\left(i k_{-2} \mathbf{n} \mathbf{r}'\right) d\mathbf{r}', \tag{57}$$

which yields, after substituting (32) and (55),

$$\mathscr{E}_1^{(0)} (\mathbf{r}) = \frac{\pi \chi_1 a_0^2 a_{-1}^*}{r} \exp\left(- i k_1 r\right) f_1 \left(\Delta k_1^{\|}\right) \left[k_{-1}^2 \sin^2 \psi - k_1^2 | \mathbf{n}^{\perp}|^2\right]^{-1} \left[k_{-1} r_0 \sin \psi J_1\left(r_0 k_{-1} \sin \psi\right)\right.$$
$$\left. \times J_0\left(k_1 r_0 | \mathbf{n}^{\perp}|\right) - k_1 r_0 | \mathbf{n}^{\perp}| J_0\left(r_0 k_{-1} \sin \psi\right) J_1\left(r_0 k_1 | \mathbf{n}^{\perp}|\right)\right], \tag{58}$$

$$\mathscr{E}_{-2}^{(0)} (\mathbf{r}) = \frac{\pi \chi_{-2} a_0^* a_{-1}^2}{r} \exp\left(- i k_{-2} r\right) f_1\left(\Delta k_{-2}^{\|}\right) \int_0^{r_0} J_0^2\left(k_{-1} r^{\perp} \sin \psi\right) J_0\left(k_{-2} r^{\perp} \sin \theta\right) r^{\perp} dr^{\perp}. \tag{59}$$

If ψ is equal to ψ_1^T, then the anti-Stokes radiation is concentrated in a narrow cone with apex angle $2\theta_1^{(1)}$ [Eq. (38)], with $\Delta k_1^{\|} = 0$ and $k_1 \sin \theta_1^{(1)} = k_{-1} \sin \psi$. The function f_1 and the second factor in (58), which describes the angular dependence, then take on the maximum possible values, namely il and $1/2$, respectively. If $\psi \neq \psi_1^T$, then the intensity of the radiation has two "flickering" maxima of $\theta_1^{(2)}$ and $\theta_1^{(3)}$ [formulas (41) and (42)], which describe the corresponding radiation cones.

Similar results are valid for the second Stokes component, as can be verified by investigating the integral factor in (59) graphically.

We note that if in addition to the field (55) with $\psi = \psi_1^T (\psi_{-2}^T)$ there is applied to the medium a beam of axial radiation of the first Stokes component, then all three types of radiation are realized for each of the components of the scattered field.

6. SPHERICAL WAVE OF FIRST STOKES COMPONENT

The case of a spherical wave $\mathscr{E}_{-1} = a_{-1} \frac{\exp\left(- i k_{-1} r\right)}{r}$ can be realized in the two-cell experiment if the first cell is a spherical volume. In addition, this case simulates the ordinary situation that arises when SRS is excited by an intense pump wave, when it can be assumed that for the greater part of the active volume the radiation of the first Stokes component, generated in its forward part, is specified in the form of a spherical wave [2]. According to (56) and (57), expanding the Stokes field in plane waves, we obtain ($B_1 = i k_{-1} r_0 \chi_1 a_0^2 a_{-1}^*$, $B_{-2} = i k_{-1} \chi_{-2} a_0^* a_{-1}^2$):

$$\mathscr{E}_1^{(0)} (\mathbf{r}) = \frac{B_1 e^{- i k_1 r}}{4\pi r} \int_0^{2\pi} \int_0^{\pi} \frac{e^{i (k_1 \cos \theta - 2 k_0 + k_{-1} \cos \theta_\sigma) l} - 1}{k_1 \cos \theta - 2 k_0 + k_{-1} \cos \theta_\sigma} \frac{J_1\left(| k_1 \mathbf{n}^{\perp} + k_{-1} \mathbf{n}_{\perp}^{\sigma} | r_0\right)}{| k_1 \mathbf{n}^{\perp} + k_{-1} \mathbf{n}_{\perp}^{\sigma} |} \sin \theta_\sigma d\theta_\sigma d\varphi_\sigma, \tag{60}$$

$$\mathscr{E}_{-2}^{(0)} (\mathbf{r}) = - \frac{B_{-2} \exp\left(i k_{-2} r\right)}{2\pi r} \int_0^{2\pi} \int_0^{\pi} \left[\int_0^l e^{i (k_0 + k_{-2} \cos \theta - 2 k_{-1} \cos \theta_\sigma) z'} dz' \int_0^{r_0} \frac{J_0\left(| k_{-2} \mathbf{n}^{\perp} - 2 k_{-1} \mathbf{n}_{\perp}^{\sigma} | r_{\perp}'\right) r_{\perp}' dr_{\perp}'}{\sqrt{r_{\perp}'^2 + z'^2}}\right] \sin \theta_\sigma d\theta_\sigma d\varphi_\sigma. \tag{61}$$

Equations (60) and (61) show that in the general case for each component there are all three types of radiation: the most intense Townes cone and two broad "flickering" maxima (Cerenkov and axial), determined by formulas (38) and (41), (42) with $\psi = 0$. We shall show that for a strongly elongated volume there will be observed only the "flickering" maxima. To this end we rewrite (60) and (61) in the equivalent form

$$\mathscr{E}_1^{(0)} (\mathbf{r}) = \frac{B_1 e^{- i k_1 r}}{2 i k_{-1} r_0 r} \int_0^l e^{i (k_1 \cos \theta - 2 k_0) z'} dz' \int_0^{r_0} \frac{e^{i k_{-1} \sqrt{r_{\perp}'^2 + z'^2}}}{2 \sqrt{r_{\perp}'^2 + z'^2}} J_0\left(k_1 | \mathbf{n}^{\perp} | r_{\perp}'\right) d\left(r_{\perp}'^2\right), \tag{60a}$$

$$\mathscr{E}_{-2}^{(0)}(\mathbf{r}) = \frac{B_{-2}\exp{(-ik_{-2}r)}}{2ik_{-1}r}\int\limits_{0}^{l} e^{i(k_0+k_{-2}\cos\theta)z'}dz'\int\limits_{0}^{r_0}\frac{e^{-2ik_{-1}\sqrt{r_\perp'^2+z'^2}\Big|_0}}{r_\perp'^2+z'^2}J_0(k_{-2}|\mathbf{n}\perp|r_\perp')\,d(r_\perp'^2). \tag{61a}$$

It is easy to show that for an elongated volume the principal contribution to the amplitudes is made by the regions where $\sqrt{r_\perp'^2 + z'^2} \approx z'$. Then from (60a) and (61a) we find immediately that the radiation is concentrated in Cerenkov cones and along the z axis (we note that consideration of the radiation of bare sources of the first Stokes component, concentrated in a strongly elongated volume [13], leads to the same results for the anti-Stokes field, if account is taken of the interference of the fields from the separate sources). The vanishing of the Townes radiation is due to the fact that the rays that are inclined at the Townes angle to the z axis leave the active volume. Thus, for strongly elongated volumes there should be no rings of class I (cf. the experiment in [4]).

The foregoing analysis shows that in the general case three types of radiation can be realized in SRS. One of them is described by the conditions of spatial synchronism (1a) for the wave equation with cubic nonlinearity. These conditions ensure a maximum energy transfer to the scattered components, which are satisfied only for a definite spatial distribution of the input fields and a special geometry of the active volume. In the opposite case, radiation of two types is produced, with intensities that depend strongly on the real geometry of the active volume. The main property of these radiations is the instability of their intensity in directions of maximum radiation relative to small changes of the geometrical parameters. We note also that this conclusion remains in force also for spatially inhomogeneous input beams.

LITERATURE CITED

1. E. Garmire, F. Pandarese, and C. H. Townes, Phys. Rev. Lett., 11:160 (1963).
2. V. N. Lugovoi, Introduction to the Theory of Stimulated Raman Scattering [in Russian], Nauka, Moscow (1968).
3. H. A. Haus, P. L. Kelley, and H. J. Zieger, Phys. Rev., 138:A960 (1965).
4. E. Garmire, Phys. Lett., 17:251 (1965).
5. V. J. Cooper and A. D. May, Appl. Phys. Lett., 7:74 (1965).
6. K. Shimoda, Jpn. J. Appl. Phys., 5:86 (1966).
7. A. N. Arbatskaya and M. M. Sushchinskii, Zh. Éksp. Teor. Fiz., 66:1993 (1974).
8. T. M. Makhviladze and M. E. Sarychev, Preprint FIAN, No. 99 (1975); Materials of the 1st All-Union Conference on Raman Spectroscopy of Light [in Russian], Znanie, Kiev (1975), p. 90.
9. T. M. Makhviladze and M. E. Sarychev, Zh. Prikl. Spektrosk., 25:1092 (1976).
10. T. M. Makhviladze and M. E. Sarychev, Preprint FIAN, No. 159 (1975).
11. T. M. Makhviladze and M. E. Sarychev, Opt. Spektrosk., 41:960 (1976); Preprint FIAN, No. 158 (1975).
12. A. Szöke, Bull. Am. Phys. Soc., 9:490 (1964).
13. V. N. Lugovoi and I. I. Sobel'man, Zh. Éksp. Teor. Fiz., 58:1283 (1970).
14. V. N. Lugovoi and A. M. Prokhorov, Zh. Éksp. Teor. Fiz., 69:84 (1975).